LIFE

THE HUMAN BEING BETWEEN LIFE AND DEATH

A Dialogue between Medicine and Philosophy:
Recurrent Issues and New Approaches

ANALECTA HUSSERLIANA

THE YEARBOOK OF PHENOMENOLOGICAL RESEARCH

VOLUME LXIV

Editor-in-Chief:

ANNA-TERESA TYMIENIECKA

The World Institute for Advanced Phenomenological Research and Learning
Belmont, Massachusetts

For sequel volumes see the end of this volume.

LIFE

THE HUMAN BEING BETWEEN LIFE AND DEATH

A Dialogue between Medicine and Philosophy: Recurrent Issues and New Approaches

Edited by

ANNA-TERESA TYMIENIECKA

The World Phenomenology Institute

and

ZBIGNIEW ZALEWSKI

The Jagiellonian University, Kraków, Poland

Published under the auspices of
The World Institute for Advanced Phenomenological Research and Learning
A-T. Tymieniecka, President

KLUWER ACADEMIC PUBLISHERS
DORDRECHT / BOSTON / LONDON

Library of Congress Cataloging-in-Publication Data is available

ISBN 0-7923-5962-3

Published by Kluwer Academic Publishers,
P. O. Box 17, 3300 AA Dordrecht, The Netherlands.

Sold and distributed in North, Central and South America
by Kluwer Academic Publishers,
101 Philip Drive, Norwell, MA 02061, U.S.A.

In all other countries, sold and distributed
by Kluwer Academic Publishers,
P.O. Box 322, 3300 AH Dordrecht, The Netherlands.

Printed on acid-free paper

Printed in the Netherlands.

TABLE OF CONTENTS

ACKNOWLEDGEMENTS

Here we bring to the public a selection of studies presented at the International Congress of Philosophy of Medicine that took place at the Jagiellonian University, Kraków, May 9–11, 1996.

This conference commemorated the seventy-fifth anniversary of the founding of a chair for the History and Philosophy of Medicine at that university. It brought together international representative medical scholars, historians, and philosophers who sought the clarification of a spectrum of issues of increasing urgency in medical science, clinical practice, and social policy.

Their focus on the human being in nature and society lends to philosophical anthropology, psychology, and sociology insights that should be welcomed there for the challenge they present to our thinking.

I am most thankful that Professor Zbigniew Zalewski, the chief organizer of the conference, has entrusted this collection to us for inclusion among the volumes on the philosophy/phenomenology of life that we are publishing in our *Analecta Husserliana* series. I believe that medicine and phenomenology of life form a natural alliance and hope that this volume will initiate closer collaboration between them.

To Isabelle Houthakker and Robert Wise go our thanks for expertly editing these papers.

A-T. T.

THE THEME

TOWARD A MEDICO-PHILOSOPHICAL ENTENTE

The art of healing has from time immemorial been at the heart of human concern. The maintenance of good health is an innermost human concern. The capacity to perform the many tasks necessitated by circumstance or prompted by desire in the course of each person's life, to accomplish the plans the human heart devises, is critical to us. But in principle we think about our medical needs only when our health is broken, when we are incapacitated and/or are suffering pain, that is, when we feel ill. We then expect that the physician will find out what is causing our illness, will locate the agents of disease or pain, and prescribe a cure or therapy.

Medical reflection and experience have developed alongside the progress of our knowledge of nature and life, and alongside our interaction with the world as we draw on the world and contribute to it. In modern times, then, medicine finds itself at a crossroads of the biological sciences and the human sciences. Yet it is raised above them by a unique concern, the life and death of the human being. Moving in the territory that stretches between conception and death, it tugs at all the existential strings that carry or obstruct the unfolding course of human life, encountering the crucial concern of every human being: his or her existence.

Here the great questions are those of the origins of life, the unfolding of the human individual, the unavoidable path to physical extinction, the question of which physical or mental condition of the individual is optimal, what is passable. Our plans, our impetus endure, but their physical platform, the forces at work in the body and its functioning, undergoes attack and suffers decline. Whatever investigations in depth can offer to revise our view of health and disease immediately interests us.

With the progress of the biological sciences, medicine became very scientifically minded, and it remains full of expectation of ever new scientific discoveries and technological inventions that it can avail itself of in maintaining people's health and curing their diseases.

While medicine and the life sciences have been forging their alliance, the practice of medicine has in our culture become ever more institutionalized and socialized. Our societies are taking on greater and greater responsibility for the health and well-being of their citizens.

Culture plays a lurking (and lingering) role in the way in which we think of health and illness, entering into the criteria of clinical decisions. More immediately evident is the role of the media in shaping expectations and opinions. Just the speed with which the world now becomes aware of new treatments, drugs, surgical procedures, therapies, and the further reach of such knowledge, is changing the ways in which physicians and patients relate to each other. Debates, controversies, legal actions, political struggles abound.

The enormous institutionalization of societal life, the laws and other measures taken to protect workers, their families, students, the unemployed, the handicapped, have extended the scope of medical reflection far into matters of social life.

A push to give the patient autonomy, to let him or her decide on treatment is undoing the traditional dependency of the patient on the physician. Personality as well as the social sphere then come into medicine's purview.

The present collection presents the panorama of the issues currently being discussed in medicine in all their practical urgency, bringing out their philosophical underpinnings. It is evident that medical reflection and bedside experience are not self-explanatory, self-sufficient. On the contrary, in every sector of theory and practice there emerge questions that refer to human nature, the human place within the unity-of-everything-that-is-alive, and human entanglements in the social network of life. These questions ask for criteria by which to appreciate situations, criteria for making judgements, making decisions, for a gradation of values, the setting of social priorities, the determination of individual rights, the weighing of autonomy versus the common good.

All such issues are by nature philosophical. Hence, the great surge of interest in our day in a philosophy of medicine. No philosophy, given its particular assumptions and biases, may do justice to these life and death issues. Yet, as much as medical reflection and theory demand philosophical insight, philosophy may in turn gain from medicine fresh insight into the nature of the human being, of society, of life. Where philosophical anthropology only deals with universals, medicine brings it face to face with difficult situations, struggle, pain, suffering, especially the suffering of not being understood.

Even as medicine refers to philosophy, each philosophical theory falls short somewhere in its approach to the world. It is the merit of the philosophy/phenomenology of life that the entire span of human life is in its scope, the whole Human Condition within the unity-of-everything-that-is-alive.

All the issues raised in this encounter of medicine and the grand sweep of life seem to converge on a chain of questions concerning the norm, the criterion, the measure for values – vital, social, and personal – when engaged in the struggle to maintain good health and vanquish or cope with illness.

It is fascinating to discover that the various philosophies of medicine not only exhibit awareness of this fact, but seek and propose answers strikingly like the main tenets of the phenomenology of life.

Thus it is that an "assay" of the two leading philosophical views on health and disease will open this collection, which points to their convergence in the phenomenology of life.

Anna-Teresa Tymieniecka

ZBIGNIEW ZALEWSKI

INTRODUCTION

A Survey of Current Issues and Approaches

Turning back to the great events of the past to celebrate them plays a relevant though mainly symbolic role in our culture. People often used to do that to reinforce their self-esteem or re-evaluate their present activities. However, the real value of such practices in our epoch, which is pragmatic and oriented toward the future, is relatively low. Going back to the great intellectual achievements of former generations of scientists or thinkers is considered by many people in practice as something similar to invoking ghosts. What can we really obtain by recollecting previous ideas and standpoints, apart from the conviction that we know more and better than our predecessors? What advantages might stem from studying former theories that may appear quite often as pretty naive oversimplifications, if not merely superstitions, in the light of our knowledge? What are, in general, the beneficial contributions to our understanding of the world that we can learn from the past?

The present collection of studies in the philosophical issues of present-day medicine is a striking example of how the urgent current concerns and debates in medical reflection as well as therapy find their ancestry in the past. It is the corroboration and often reformulation of these issues which fascinates both the medical scientist and the philosopher.

The papers in this book are a selection of those presented at the International Conference held at the Jagiellonian University, Kraków on May 9–11, 1996, commemorating the 75th anniversary of the founding of the chair of History and Philosophy of Medicine at this University. Among the great issues of present-day discussion, several are brought back to the original form in which they had been raised by eminent Polish scholars. This historical perspective gives additional depth and insights to the contemporary approaches.[1]

Looking at the debates carried on by Polish physician–philosophers nearly a century ago, we find an answer to the questions raised. One finds there quite familiar topics and approaches. The origin and growth of scientific knowledge in medicine, ways of finding diagnostic and therapeutic indications, methods of reasoning and understanding in medicine, the place of knowledge and experience in medical endeavors, the methodological status of medicine as a science and art of healing at the same time, and moral quandaries stemming from everyday medical practice, are among the most

A.-T. Tymieniecka and Z. Zalewski (eds.), Analecta Husserliana LXIV, 1–11.

relevant issues which were intensively discussed then and are still discussed nowadays. However, not only were these topics discussed a century ago, but one can also realize that the answers given to the questions undertaken at that time are interesting and valid for related contemporary controversies. And the approach to medicine that was revealed at that time, such as the tendency to think about medicine from the bedside rather than from the summits of academic posts, while being an overt postulate in modern times, was undoubtedly unique and insightful.[2]

To find a vivid tradition resembling our present-day experience is merely the first step leading to its appreciation. The discovery of the relevance of thoughts from the past would not necessarily entail their approval by modern people or their inclusion in the debates over current controversies. And yet there are plenty of deep, wise and momentous ideas in the intellectual treasury of humankind, to which contemporary people are completely insensitive. Hence, there must be something either in the evoked historical period, or in a comprehension of the reality people are living in, which causes them to refer to their ancestors' ideas in order to seek inspiration for resolving their own problems. At least three domains of the medical field cultivated by Polish philosophers seem to be particularly interesting and valid in our own work of coping with medicine in the modern world.

Critical appraisal of scientific knowledge and its growth in medicine is one of the outputs of the Polish School, which can hardly be overestimated. Worshipping the sciences as such, a process which emerged in the nineteenth century but expanded to an enormous extent within our era, has confronted us with the problem of justifying the real value of scientific development. Seduced by the extraordinary progress of medical technology and fascinated by a gigantic increase of information, the modern physician is very often disorientated, and left with a scarcity of means and tools for comprehending the acquired data. The accompanying pressure of being up-to-date with the newest discoveries and achievements of medical sciences makes him accept any novelties that come along with a "trademark" of science, i.e., everything that appears in the latest issue of one of the respectable medical journals. Left alone in the face of all of this, the physician is quite often unable to understand exactly and evaluate properly what is presented as knowledge. Even worse, he seldom feels a need to do so. The situation described belongs to the most significant symptoms of the dehumanization syndrome, which is characteristic of modern medicine. That is why the ability to take a critical approach to and a rational distance from excessive demands to be scientific, together with an open-mindedness and willingness to approve and apply any valuable innovations to one's own practice, are essential elements in the education of future healers, and among those currently practicing.

Looking at the patient through "scientific spectacles" allows a modern physician to discern failures and defects of different organs or tissues within the human body, and even to perceive malfunctions or deformities of individual cells. However, it does not allow him to become aware of the very nature of the entity standing face-to-face with him. An ill and suffering human being as such remains outside the scope of interest of a scientifically equipped physician. Everything but the encounter with a sick and deteriorated fellow person falls among the numerous tasks imposed on the physician by medical enterprise. The structure of scientific knowledge directs his attention to the diseases which, from the theoretically elaborated, scientific point of view, "are internal states that depress a functional ability below species-typical levels, [...] i.e., reduce one or more functional abilities below typical efficiency".[3] Defined in such a way and strictly described in medical textbooks, diseases should be diagnosed and cured, which means: recognized and fought. These aims empower the physician to use a myriad of technical devices and gadgets in order to be effective in recognizing these "reduced functional abilities" and coping with them. Disease is transformed into a "disease unit" and is taken as an enemy, while the patient is nothing but a battlefield. Since the plea for objectivity ascribed to scientific medical knowledge distances disease from those who are diseased, medicine itself becomes an entirely depersonalized enterprise. Instead of being personal, the relationship between the physician and the patient is merely technical and devoid of sensitivity and compassion. However, the physician is only partly a chaplain in the temple of science. In the other part of his self, in his other embodiment, he is a healer dealing with the real suffering of real people. Yet those two roles of the modern physician are visibly incompatible with each other, which throws him into a kind of cognitive dissonance, if not into a split personality. Consequently, this incompatibility prevents him from recognizing the patient as a person. The latter remains an unknown entity to his healer. How to bridge that gap between scientific knowledge and the need for human feelings, how to enable a modern physician to reestablish humane relations with his patients and to see them, above all, as human beings – how to, roughly speaking, rehumanize modern medicine – is the greatest challenge to our conception of medicine and its functions on the threshold of the next millennium. The lessons of medicine practiced from the bedside, stemming from the past, might be very instructive for us.

Moral discourse has dominated the development of the philosophy of medicine in the last quarter of the twentieth century. Its enormous increase, as well as the fact that it has gained very broad public reception, is due to the growing relevance attributed to moral and legal aspects of controversies arising in the progress of biomedical sciences and their technical and technological utilization in medical practice. It is this emphasis placed on the moral dimensions of healing practices that leads to the use of the word "bioethics" as a proper name for that domain of intellectual activity. The time pressure stemming from the urgent need for immediate solutions to practical problems emerging in everyday medical proceedings has caused bioethics to fall rapidly into a state which might be called "a crisis of foundations".[4] None of the proposed ways of approaching and answering moral queries proved satisfactory. None of them could lead to consensus in coping with controversies and establishing universally valid standards for solving moral dilemmas. Hence, none of them might appear as a handy tool for everyday use. Bioethics is also stigmatized by the moral diversity of modern pluralistic societies. Moreover, haste caused by a demand for quick answers to compelling questions and the requirement to deliver a universal algorithm for solving problems also marked bioethics in another way. Bioethical textbooks are full of very meticulous analyses of particular practical problems or cases approached from different angles and depicted in various aspects, but only a few of them contain more than an outline of general ethical quandaries. The abundance of practical deliberations is contrasted here with the scarcity of theoretical moral inquiries, so that solutions provided in such a way seem to be hung out in an intellectual vacuum. Integral approaches to human persons and their affairs disappear, being replaced by a mosaic of detailed items. Hence, the attempts to reunify human understanding of the surrounding world as a whole, and to perceive each problem in its reciprocal relationships, are so important for contemporary people. Looking back to the heritage of past, thinkers can help us deal with this task. Nonetheless, our turning toward tradition does not simply mean imitating solutions taken from the past and applying them to our problems. It should rather follow the appraisal we refer to. It can encourage us to rethink our own matters, including moral matters, and to find our own adequate solutions to the present state of affairs in a more comprehensive way.

Modern medicine, with its enormous progress, poses a huge challenge to our thinking about ourselves and the world we are living in. And not only can it save the life of ethics, as S. Toulmin stated in a somewhat pompous manner,[5] but it can also give a strong invigorating impulse to all philosophical inquiries. Scientific progress has caused many of our essential convictions regarding man himself and his position among other creatures in the world to become doubtful, if not evidently false. By transcending the borders between life and death through the excessive use of life-support machinery and the artificial creation of new human beings, by transplanting hearts, kidneys, livers etc., and by manipulating the human genetic heritage – those being the most eloquent examples – contemporary medicine demands that we shift our attention to its affairs. As its rapid progress considerably changes our lives and will give rise to further, much more far-reaching alterations, our thinking should follow these metamorphoses. We should again ask each of the great metaphysical, epistemological and moral questions and painstakingly try to find adequate answers to them, formulating our intellectual investigations from within the perspective structured by medicine. This is, or might be, the most suitable way to deal with the mental chaos of postmodernity and to try to overcome it, and this is, as well, the best way to pay homage to the dignified tradition.

The collection of essays presented here is an initial attempt of this kind. Referring directly or indirectly to the intellectual legacy of the Polish School of Philosophy of Medicine, the authors from various countries try to rethink a cluster of important issues posed by modern medicine, tracing the general patterns of thought established in the invoked tradition. Three most important domains of interest, which are mentioned above, are the leading subject matter of their considerations. Attention has shifted to the logical and methodological aspects of medical knowledge, and various approaches to the cardinal medical concepts of health and disease remind us that this area of intellectual explorations was a field of the greatest interest within the Polish School. As to our motto, we can take the title given to that part: "The Logic of Medicine or the Critique of Medical Understanding",[6] which is also the title of probably the most mature book (besides the aforementioned work of L. Fleck) in the output of the School. The book, quoted above, was written by Władysław Biegański, one of the leading members of the School.

Changing images of the human being in current medicine and contemporary philosophy are analyzed in the second part of the collection. Not only do the authors penetrate the different dimensions in which man is contemplated in the circle of European thought, but also, in their search for a deeper understanding of modern man, they break the limits established by West-centered culture, and try to find inspiration in Jewish and far-Eastern traditions. Finally, moral dilemmas put forward by modern medicine are dealt with. The value of life, the person's autonomy and the principle of justice are among other subjects discussed both in general and with regard to particular problems posed by everyday medical procedures.[7]

The set of problems covered under the name "Logic of Medicine" is undoubtedly of great value to current philosophical debates over medicine and may claim its ancestry in the output of the Polish School. Questions regarding the methodological status of medicine are crucial for the discernment of its place and role in society. Medicine is a complex venture which embraces knowing and acting, encompassing human beings as its subjects. Therefore, medical knowledge, contrary to other, so-called "pure sciences", must inevitably be descriptive and evaluative at the same time. Any efforts to separate explanation from evaluation might thus be effective only to a very limited extent. Nonetheless, many biomedical scientists in our century, inspired by positivistic ideas concerning pure descriptiveness as *conditio sine qua non* of science, have tried to establish and develop medical sciences imitating patterns taken from other natural sciences. This approach, implemented into medical education and training for years, has structured the attitudes of a vast sector of the medical profession. It, however, has necessarily led to the reduction of manifold human phenomena to simpler processes occurring merely on the organic level, while mental and spiritual aspects of human life have been exorcised from the domain of science. For medicine, man has become nothing but the recipient of an organic dysfunction called disease, while illnesses or suffering occur as something that medicine has nothing to do with.

The rediscovery of the patient as a person appeared in contemporary discussions about medicine some thirty years ago.[8] Biomedical models of medicine as science, and of man as its subject, have been challenged. New areas of human life (psychical, societal, environmental, and so on) have been included in the scope of medical interest. Values, previously distorted from the scientific realm of medicine, returned and affected the way man and his problems were comprehended. The myth of medical knowledge as free from evaluation vanished. Thus the account of medicine needed radical changes with this reappearance of values in the realm of pure science. The new model had to provide room for an axiological dimension of human life, and had to include human values and goals in its considerations (Zbigniew Szawarski), while not expunging purely biological knowledge of somatic processes, but setting the limits of its validity within the wider framework of understanding humanity. No longer could the logic of medical reasoning be exclusively the logic of discovery and explanation, it also had to be the logic of evaluation (Kevin Wildes). The coexistence of those different models and different forms of logic within them creates medicine as a unique type of knowledge with complex methodological patterns and complicated networks of mutual interrelations between them. Hence the progress of medical knowledge depends on developing modes and models of reasoning and proceeding in each dimension, while staying in touch with, or at least keeping in mind, requirements stemming from the others. Thus the clear logical analysis by Jan Doroszewski, of basic empirical knowledge on a biomedical level, consisting of declarative (i.e. descriptive), operational and instrumental knowledge, which may be helpful in avoiding mistakes in everyday medical procedures, cannot be overestimated. Scientific medicine, however, is not the only type of healing practice currently existing in our culture. There are plenty of other practices based on different systems of healing knowledge and founded on different images of man. The significance of these alternative practices for medicine and philosophical reflection about medicine, as well as the reasons why so many people prefer those alternate practices to conventional medicine nowadays, are explored by Joachim Widder.

Health, disease and illness are among the basic themes of interest to those who focus their philosophical attention on medical matters. Debates concerning the meaning of these concepts have enormous longevity and recur in every epoch. One can even consider them as never-ending stories, especially as regards the recurrence of certain motifs discussed by Chałubiński *et al.* a century ago, and compare them to the famous controversy of the 1970s and 1980s. Two authors in this volume attempt to participate in this controversy. Piotr Mróz approaches mental disorders phenomenologically, finding inspiration in the existential phenomenology of Jean-Paul Sartre and Maurice Merleau-Ponty. Jan Hartman investigates the evils and goods of illness, visibly transcending the borders between the two forms of medical logic previously disclosed.

Another main topic of thought concerning medicine, namely the question of the relationship between medicine and art, is dealt with in two essays written by aestheticians. Maria Gołaszewska draws an outline of an aesthetics of medicine as it has been reflected in beaux arts since ancient Greece to modern literature and mass-culture productions (TV series). Grazẏna Podraza-Ucińska presents an attentive analysis of different meanings in which a notion of art can be applied to medical practice. Stressing the role of creativity, intuition (in various senses) and practical skills in the actions performed by doctors, she emphasizes different aspects of similarity between the art of healing and the creation of works of art.

Although the questions about medicine as such and a bunch of issues stemming from its intellectual elaboration are a matter of enormous relevance for a rational approach to the phenomena of health and disease, the human being himself in his manifold manifestations remains a central problem of each act of truly philosophical reflection. Whatever the subject of analysis might be, whether it be science or common-sense knowledge, medicine or art, ways of thinking inevitably lead to a human person, their bearer, and to different roles that a human person plays with regard to them. The second part of the book is devoted to the question, Kantian in its very essence, of "who a man is", and to the various ways in which this question is posed in contemporary philosophy, while the answers take into account the intellectual stimulation of modern medicine.

Recent philosophy, which frequently calls itself "post-modern philosophy", looks as if it had lost the solid ground of the universally valid ideals of truth, wisdom and rationality. A state of uncertainty and mental confusion, due to latest fashion trends in thinking (like deconstructionism or similar intellectual novelties), has resulted in a disintegration of the realm of philosophy. Instead of following strict rules of thought and analysis, philosophers quite often devote their inventiveness to multiplying the ways in which the modern *conditio humana*, is approached, causing a growth of confusion rather than a growth of knowledge and understanding of the matter. Thus, one is faced with a wide variety of philosophical beliefs, which are often nothing but the expression of someone's convictions which are not necessarily well grounded or sufficiently justified. An omnipotent ghost of relativism is cruising over Western culture. Changing or "oscillating" images of man in contemporary philosophy and their validity for the medical comprehension of human beings are the subjects of deliberate study by Henk ten Have. Yet it is not only the development of philosophical thought that affects the vision of man in modern times. The rapid progress of biomedical sciences, also the concept of the human body, subjects resulting from that transformation, are carefully analyzed by Christian Byk. As the Western intellectual heritage seems to be too narrow for many thinkers nowadays, they reach out for the legacy of Eastern cultures. The human body in its relation to the whole human entity, as it is understood in Hindu and Yoga traditions, is the topic of the next two essays, written by Marta Kudelska and Marzenna Jakubczak.

Values were invented in twentieth-century philosophical anthropology as indispensably involved with human life, to such an extent that humanity would be incomprehensible without them. They pervade each dimension of human life. They structure our imagination, mold our cognition, set a hierarchy of goals, affect and organize our activity while directing it to achieve those goals, and develop our sensitivity. They cause us to appreciate life as worth living, and health as worth preserving. Life and health taken as values can, however, be in collision. Considered on a deep existential level, as is done by Kornel Gibiński, they may lead to a radical dilemma: to have (good health) or to be (alive). Man confronted with death is the subject of Kazimierz Szewczyk's diligent inquiry, and existential values in general preoccupy Józef Lipiec. Personalism as a view of man as a moral entity from its very nature, and a particular type of personalism in its possible application to bioethics, are analyzed by Maurice de Wachter.

The problem of values and their impact on human life leads to the last part of the book, which is devoted to deliberate studies concerning moral dilemmas indicated by modern medicine. The value of life and the person's autonomy are the leitmotifs of these studies.

H. Tristram Engelhardt, Jr. confronts changing attitudes towards the borders of life in modern society (taking into account abortion and euthanasia), with the crisis of Christianity, the milestone of Western culture. Robert Barnet analyzes changing values against the background of the megatrends undermining the development of modern medicine, which are medicalization, technologization and commercialization.

A clash between paternalism and autonomy is another of the most resounding subjects for heated discussions in bioethics. Donald Evans investigates how the concept of impaired capacity is constructed and how is it used in cases of children and mentally disabled patients to legitimize procedures for making clinical decisions in surrogacy. Hans-Martin Sass explores the problem of who is entitled to make decisions regarding persons in terminal states and votes for advance directives, expressed in the form of a Living Will, as the ultimately moral way of solving that question. Other aspects of making decisions on behalf of a patient are discussed in the essays of Barbara Maroszyńska-Jezòwska and Jacek Jaśtal. Anna Alichniewicz approaches the doctor-patient relationship from another angle, asking about the limits of paternalism and autonomy and about a balance between justice and beneficence in cases of scarcity of health-care resources. Finally, Ignacy Fiut follows the way in which the value of health is reflected in the Polish media.

Two essays closing the collection are devoted to the "lessons from the past". Ilana Löwy stresses essential features of thought developed by eminent members of the Polish School. By taking into account historical, socio-cultural and methodological contexts of scientific knowledge and linking theoretical considerations with concrete circumstances bound to particular cases taken from regular practice, those doctors were able to comprehend the patient as a person in his integrity, and not merely as an interesting exemplification of a certain "disease unit". Analyzing the idea of a humanistic approach, called "philosophizing at the bedside", which might be held as a hallmark of the message from the past addressed to modern medicine, Józef Kałuzà shifts his attention directly to Władysław Szumowski, the founder of the Chair of History and Philosophy of Medicine at the Jagiellonian University, emphasizing his great interest in the logic of medical reasoning and related issues.

Department of Philosophy of Medicine, Jagiellonian University, Kraków

NOTES

[1] It was in the academic year 1920/21 that the first chair in the world with the term "philosophy of medicine" in its name was founded at the Jagiellonian University in Kraków; and later in that same year, similar chairs were established at the other Polish universities. This unique stream of intellectual inquiries, called the *Polish School of Philosophy of Medicine*, which had been developing within a large group of Polish philosophizing physicians since the 1870s, finally gained institutional nobility and thus the philosophy of medicine for the first time achieved the status of an academic domain. Unfortunately, the existence of both the School and the Chairs was cut short by World War II, an after the war's end it turned out that there was no place for humanities in medical education in the post-war reality dominated by Communism. Chairs were dissolved, and philosophical reflection about medicine survived only as a kind of private hobby among some open-minded physicians. The important output of that School was thus forgotten for years in Poland, and for years it remained unknown outside the country's boundaries.

At the same time, interest in the problems and questions undertaken within the Polish School was growing in the western world, so in the early 1970s a new academic discipline usually called "bioethics" or, somewhat less frequently, "philosophy of medicine" was formed. Although bioethics expanded immediately after its birth and rapidly became a powerful and influential domain of philosophical research, the interest concerning the works of its Polish ancestors, which started with the discovery of Ludwick Fleck's work by Thomas Kuhn, was growing much more slowly and in a rather limited scope. Therefore the intellectual achievements of Polish pre-war philosophy of medicine became better known all over the world only when the book *The Polish School of Philosophy of Medicine* by Ilana Löwy was published in 1990.

Kuhn mentioned L. Fleck's book *Genesis and Development of a Scientific Fact* in the introduction to his *The Structure of Scientific Revolutions*. Growing interest in Fleck's ideas induced by Kuhn's remark turned the attention of many philosophers of science to the origin and intellectual background of his work and resulted in the discovery of the Polish School by western philosophy (it needed discovery because almost all of its works were printed exclusively in Polish; Fleck's case of a German edition was almost exceptional). Cf. also Ilana Löwy, *The Polish School of Philosophy of Medicine. From Tytus Chałubiński (1820–1889) to Ludwick Fleck (1896–1961)*. This book contains not only essays devoted to the main members of the school mentioned, but also a representative selection of their works.

[2] Biegański, W., *Logika medycyny, albo krytyka poznania lekarskiego (Logic of medicine or the critique of medical understanding)*: Warsaw: E. Wende, 1908 (German edition: Biegański, W., 1909, *Medizinische Logik. Kritik der Ärztlichen Erkenntinis.* Würzburg: Kabitzsch, 1909).
[3] Boorse, C., "Health as a Theoretical Concept". *Philosophy of Science* (1977), 44, 542–573.
[4] The state of modern bioethics is broadly analyzed, for instance, by H. T. Engelhardt, Jr. in his *The Foundations of Bioethics*, New York, Oxford University Press, 1996, 2nd edition (see especially chapter 2: "The Intellectual Bases of Bioethics", pp. 32–101).
[5] Toulmin, S., "How Medicine Saved the Life of Ethics", *Perspectives in Biology and Medicine* (1982), 25, 736–750.
[6] Biegański, *op. cit.*
[7] As it is obviously impossible to sharply separate these three domains of reflection in particular texts, similar topics are presented in different parts of the book. Nonetheless, a qualification is made according to the main subjects of analyses.
[8] A book which had symbolic meaning for that tendency (entitled *The Patient as Person*, New Haven, Yale University Press, 1970), written by Paul Ramsey, came out in 1970.

ANNA-TERESA TYMIENIECKA

THE ONTOPOIETIC DESIGN OF LIFE AND MEDICINE'S SEARCH FOR THE NORM

An Alliance between Phenomenology/Philosophy of Life and Medicine

INTRODUCTION

1. *The Issues*

The Romantic philosopher Schelling, who showed profound interest in medicine and whose philosophy attracted considerable interest in the medical profession of Germany in his day, pointed out that medicine is the "royal road" of philosophy. This insight has come to full fruition in the contemporary phenomenology/philosophy of life.[1]

Both phenomenology of life and medicine are focused directly on life. Furthermore, just as the phenomenology/philosophy of life looks to medicine for concrete insights into the nature of life in all its byways, so does medicine find in the phenomenology of life a legitimate expansion of its reflections and an encompassing ultimate (final?) frame of reference for innumerable threads of experience that would otherwise remain but loosely connected. In the present study I intend to offer a succinct presentation of this alliance.

The essential interests of these two enterprises span the two radical frontiers of each life's beginning and its termination in death. Between those limits lie the vast intermediary zones of subterranean transformatory networks, the womb of nature, out of which emerge the significant moves of individual existence. Both medicine and phenomenology of life seek there the explanation of the forces, organization, and order to be found in the hurtling rush of life. They necessarily engage all the branches of science that tie in with the science of nature-life: astrophysics, particle physics, magnetism, electrokinetics, chemistry, metallurgy, meteorology....

But if medical science and practice is to bring theoretical enlightenment to and serve individual life it must engage the philosophy/phenomenology of life in a common exploration of the world of life, the societal world, and the world of human culture. In dealing with individual persons' well-being or illness/injury, the healer has to see them in their interactions with those worlds. Medical science itself is impelled and oriented not so much by scientific research and technological invention as by the great trends in

A.-T. Tymieniecka and Z. Zalewski (eds.), Analecta Husserliana LXIV, 13–36.

society's preoccupations, by style and culture. And today, as never before in the history of Western culture, medicine is immersed in social debate.

While there are academic debates about just how much "progress" has actually been accomplished in medicine in our century,[2] the unprecedented pace of technological invention has greatly transformed medical practice by both prolonging life and bettering its "quality," even for those incapacitated. Further innovation is opening the way to transforming the "innate" human endowment, whether organic or psychic.

On the one hand, philosophy of life follows the scientific tunnelling into the subterranean moves of life and watches in fascination the ways in which medicine applies the findings in individual applications. Investigation and service to life meet.[3]

On the other hand, medicine looks up to philosophy in several ways. First, healers realize that the innovations in science and medicine are occurring at a rate beyond which human beings can properly absorb and evaluate change. This is especially so when we consider that in our times the moral-aesthetic-religious threads of the cultural fabric that has heretofore inspired and sustained personal and social evaluation, expressed the very meaning of the individual and the sense of social life, is dissolving.

Principles, resources, criteria of old do not hold and new ones are missing. The decisions of physicians culminate in those of life and death. As ethically and religiously derived codes for behavior fade, healers find themselves in a conundrum of perplexities as they weigh whether to act or not to act.[4] Each decision draws on numerous perspectives, attitudes. Not only is the right diagnosis sought, but also the appropriate attitude to be taken in treatment. But, as I said, the principles by which consensus may be reached are not available. Even the question of the norm for health and disease, which lies at the heart of medicine, is being debated.

2. *Corroboration of the Issues via a Historical Digression*

From time immemorial medicine, or the art of healing, has always served the same purpose: maintaining health and preserving life. But the concepts of health and disease have changed in the succession of history's civilizations and are still debated. From what we learn of him through Aetius, Alcmaeon of Croton, a Pythagorean of the last third of the sixth century B.C. and the first Greek medical man we have record of, established a basic framework for medical reflection. First of all, he held the difference between health and illness to be determined by the proportions of the forces at play in the body.

Health consists in an equilibrium among these forces, disease in the domination by one of them. Second, he distinguished between internal causes of disequilibrium stemming from the body itself and external causes coming from the environment. Third, he differentiated between the physiological functioning manifested in the interplay of bodily organs and the psychic life he saw emanating from the brain. (Therein he saw the difference between human beings and animals.)

And so a schema was established for medicine. The notion of the forces comes up again and again. There are the four "humors" whose proportions Hippocrates saw as fundamental to health and illness. There is Paracelsus' notion of the vital force, which he called the "archaeus." With John Brown the "archaeus" became an "incitation" initiating life. For Schelling sensibility and irritability were two distinct forces distributed throughout the organism.[5]

In the scenario of forces, the key issue for health and disease was that of the proportions of the forces in their distribution as well as their enactment/usage. The norm always held up was that of equilibrium (homeostasis).

The more detailed analytic differentiation of the functions of life achieved in the seventeenth century after the invention of the microscope allowed the functioning of an organism's unfolding and maintenance of life to be better apprehended and challenged the foundations of medical thinking.

It was Rudolf Virchow's work that allowed a radical breakthrough in perspectives.[6] Going down to the cellular level of the organism, he located the primordial play of forces there. It is out of cell-level concatenations that the organization and functioning of the individual body/life proceeds.

Virchow's primordiality of the cell in the constructivism of life has now in turn been fundamentally qualified by more elementary findings about life's components. While at the cellular level the constructivism of the individualizing process is to be seen, contemporary genetics reveals the pre-constructive workings of life's pre-elements as they perform complex subsidiary, subjacent elementary operations.

Still the cell remains the crux of the constructivism of life. What is fascinating about the cell is that although we can say that it is the building block of individualizing life, it is not a mere "atom" of life. On the contrary, it already exhibits the marks of a singular living being, namely, self-maintenance and self-replication. Virchow emphasizes that the organism as a whole is a society of innumerable societies of cells maintaining functional connections unfathomed in their number and variety.[7] They form a functional network in which the forces of life intergenerate, weaken, and vanish with the emergence of a progressively harmonized living system, what we call an "autonomous living being."

The cell offers a platform, then, from which we may, in one direction, follow the construction of life's constructive arteries and, in the other direction, probe its preconstructive operational segments.

Recent genetic research is reaching still deeper levels of life's organization. Direct and precise interventions triggering the constructive capacities of genes allow us to become "genetic engineers." The prospect of being able to counter hereditary dispositions to disease at that level fills medical science with great hopes for further discoveries. To each physiological defect there may be found a genetic key. The mapping of the entire genetic substructure of the human being, the human genome, is much anticipated.

But medical practice also needs a succinct outline of how individuals function in their external environments. As Virchow emphasized, we cannot fully understand the role of a part without considering it within the whole it is part of.[8] To offer a map, a plan of the "whole" is the role of the phenomenology of life that we are unfolding.

As we further differentiate the levels of the play of life's forces, we have to seek the measure of health and disease in the universal plan of life's individualizing organization also.[9]

In fact, as I noted before, the norm for judging health and disease is disputed nowadays in medical circles.[10]

3. *The Debate on Health and Disease*

Today's medical reflections and discussion, whether they be in the realm of theory or that of therapy, take in the whole scope of human existence. Under the purview of medicine lies everything from the preconstructive genetic material to the complexities subjacent to life that it forms, to the nature and functioning of the organism, to the brain that is the center to which all the threads of life tie in and is considered the site of the specifically human interpretation of life, to our social and cultural life, all human doing and creation.

Discussion of the definitions of health and disease focuses on the question of what is the norm, the norm for diagnosis, for prescribing treatment. A refinement is the issue of on what basis to legitimate the decision to provide a health promoting medical device to an individual and the issue of whether the criteria guiding the decision should be set in the public sector. In short, both the norms for judging situations and for making decisions are in question.

Generally speaking, two basic perspectives on the notions of "health" and "disease" predominate in the contemporary discussion. One of them focuses on the human organism and analyzes its structure and functioning. Those who hold it maintain that we may arrive at the norm of health and disease through biological analysis alone.[11] The other perspective takes into consideration the entire human being and that person as he or she is immersed in the conundrum of an interaction with the world that is taking or attempting to take its own course.[12] This "holistic" approach takes its cues mostly within the sphere of human sociocultural involvement and draws on the findings of sociology and anthropology in determining what is health.

Although any complete and sound theory of health should consider both perspectives, they do not necessarily coincide in their determinations of what is basic. They may envisage normalcy with reference to different basic data. The biological perspective claims that only biological principles of evaluation bear on the question. In the holistic perspective, in radical contrast, ethical, religious, aesthetic, and purely personal convictions are given weight, the life directives of the entire person by which he establishes his very own significant universe of life and according to which he finds meaning and happiness.

What is at stake here is the "nature" of the human being, and that is naturally of paramount importance for philosophy. This is a philosophical question, one at the heart of phenomenology/philosophy of life and its elucidations.

4. *The Ground for Our Assay of Perspectives*

Medical science appears to be acutely aware of its ties to philosophy. Philosophy of medicine, to judge by the numerous conferences, journals, and books devoted to it, is flourishing. Medicine is looking to philosophy for help in disentangling the complex issues facing it. Medicine is drawing on all the many fields in which questions concerning human beings are researched, but needs universal ideas that tie together all the particular questions. The questions by themselves hang loose and beg for continuation at a level higher than the empirical and broader than strictly medical theory.

Philosophy, especially the phenomenology of life, focuses on the same issues of life and death, being an enterprise of universal research, one that goes into all the sectors of existence in which medicine moves. Its impartial witness to all that is makes it a unique partner for medicine in its search for final, overarching answers. The present-day cry for guidance in handling the practical issues of life and death make the alliance between medicine and the phenomenology of life a most significant event.[13]

It is against this just sketched background covering the two main perspectives discussed in medicine today that I will present two medico-philosophical theories that are representative of them. First, I will review Christopher Boorse's biologico-analytical theory of health and disease with its reductionistic tendencies.[14] Then I will turn to Lennart Nordenfelt's theory of health and illness, one representative of the holistic perspective, one that aims to draw on the full expansion of the human life enactment, and also one that – although taking into account all "humanistic" concerns – tends to draw on sociological intuitions and theories especially, clearly limiting its philosophico-analytical scope.[15]

Each of these almost radically divergent theories, which have received a great deal of attention in contemporary philosophy of medicine, crucially coincide with the phenomenology of life's guiding concept, that of life's self-individualizing progress. The confrontation of phenomenology of life with these two perspectives will yield a host of ontological-ontopoietic insights to which they already point. We will see how the concept of the ontopoiesis of life offers to these radically approaches to "human nature" the arch toward which they reach.

PART ONE
THE BIOLOGICO-ORGANIC PLATFORM

1. *Boorse's Statistical Approach as a Representative Perspective*

Christopher Boorse's analysis has, as he points out, an underlying intuition that health is the normal condition of the human being, which, we may say, has been the intuition of the medical tradition. He assumes that "normal" is "natural"[16] and that deviations from the normal constitute illness. Stating that, he immediately strikes the nail on the head, saying that the "normality" of health resides in the internal functioning of the organism. In this he refers to a historical survey by Temkin[17] that Boorse believes finds the same viewpoint all the way back to Galen. But he also refers to similar contemporary views such as those of King.[18]

This viewpoint belongs with a host of teleological and typological notions having Aristotelian origins. Boorse's definition of health as the absence of disease relies on two root notions: biological functioning and statistical normality.[19] Each function is goal-oriented. Boorse offers a brief account of a functional network in which each organ takes on its given role and works together with the others in the whole body schema. The functional network of the organism comprises all the types of operations, processes, etc. that take place within it. They are hierarchically arranged in a way that is conducive to their carrying out together the growth and development of the living individual. Within this hierarchy there is an intrinsic principle determining the primary function that each organ performs. We may say that it is not the "system" that is being served, but that principle. For example, the kidney's role is to "eliminate waste rather than to keep the bladder full."

The organism brings together the performance of a bundle of simultaneous functions. Different sciences focus on different ones of these, for example, on individual survival, continuation of the species, ecological equilibrium with the environment or its disruption. Emphasis falls, however, on the core functioning, which is a fundamental concept reposing on itself. Statements and judgements that concern functions directly always repose in themselves and are aloof from ethical or social evaluation.

Together with the notion of functioning goes that of statistical normalcy. Although Boorse notes that statistics yield neither sufficient nor certain results, it is the best way in which medicine may differentiate "natural classes" of organisms and so produce "ideal types." The ideal type constitutes then a "reference class."

And so it is that at this point Boorse, in order to make the distinction between health and disease upon his platform of biological functionalism, introduces the striking notion of "species design."[20]

This functional conception of Boorse's would encompass the entire system, horizontally and vertically. A further qualification of physiological functioning is required. Since physiological functions are meant to be "standard contributions" to the survival and reproductive performance of the given individual, we need to establish what is "standard" first. And then this standard has to apply to a "population" or reference class to which the individual in question belongs, that is, to a species.[21]

The notion of species design serves to indicate the nature of the "reference class," which then may be defined as "a natural class of organisms of uniform functional design, specifically of an age group of a sex of a species."[22]

Boorse's asserted aim in introducing the concept of species design is in his words "to analyze health in its conformity to it." In short, species design is the standard of normal functioning. Normal functioning "in a member of the reference class is the performance by each internal part of all its statistically typical functions with at least statistically typical efficiency, i.e., an efficiency level within or above some chosen central region of their population distribution."[23]

With reference to this class, the "normal function of a part or of a process" is conceived as being the "statistically typical contribution by it to their individual survival and reproduction."[24] And so health is defined as *"normal functional ability: the readiness of each internal part to perform all its normal functions on typical occasions with at least typical efficiency"* (emphasis added). Disease, then, is any *"internal state" that "imperils health"* (emphasis added).[25] Boorse affirms that in this understanding, "The history of medical theory is nothing but a record of progressive investigation of normal functioning on the organismic, organic, histological, cellular, biochemical levels of organisation, and of the increasingly subtle kinds of pathology this investigation reveals."[26]

This is in fact a notion made up of two amalgamated concepts: the concept of a class of living beings, such as humans, that serves as a frame reference for the other concept, which is strictly and essentially correlated with the first, that of the definition of physiological functioning using a universal notion of a "standard" type of functioning. With these two correlated concepts in place, a differentiating principle is introduced. Insofar as "function statements" do not describe an individual's functioning, they "describe species or populations characteristics."[27] A species, apprehended according to the peculiar features of the organism, is an "ideal type of organism."[28]

The concept of species is considered the foundation of comparative physiology. Boorse refers to the typical textbook physiology in which the various types of living beings are distinguished and in which, "For each type a textbook provides a composite portrait of what I call the *species design*, i.e., the typical hierarchy of interlocking functional systems that supports the life of an organism of that type."[29] The significant point to be emphasized is that this "idealization is of course statistical," which amounts to saying it is a generalization from a set of individual instances. Thus, the differentiation of species as well as of species designs is an "idealization" of a statistical sort. We will return to this later. Worth mentioning now, is that this type of idealization guarantees, in Boorse's view, its strictly biological, matter of fact nature, allowing no admixture of considerations of an aesthetic, moral, or generally "normative" nature.

Boorse will not concede to Tristram Engelhardt that considerations such as "suffering, goals proper to humans, . . . the form and appearance proper to humans" shape our complex judgements about health or disease. He insists that all of Engelhardt's examples basically involve "failure of parts of the body to perform biological functions which it is statistically normal for them to perform."[30]

As a matter of fact, one reason for which this serious effort by Boorse is, and should be, awakening great interest is that today medical issues are more and more entangled in social considerations, in the personal considerations of the patient, in the theoretical biases of the healer, in societal policies on preferred treatment and access to it. It would be most significant if we could refer the question of health and disease to an objective norm not relative to social concerns, to physiology itself.

If such an objective universal principle framing health and disease in terms of normality and deviation from it could be established, what guidance there would be for medicine in it! What is at question is whether the biologico-physiological norm – as basic as it is – could effectively serve as such an objective referential schema.

2. *Biological Species Design and the Ontopoietic Design of the Tree of Life*

Boorse's biological concept of species design coincides strikingly with the phenomenology of life's concept of the ontopoietic design of life. The two approaches, one focusing on the singular organism as a representative of a species, the other on self-individualizing life, appear to stem from two frameworks that are congenial, the discussion of which we now begin.

First of all, the biological theory of health and phenomenology of life share the same basic platform, namely, *the functional system of the unfolding and progress of life*. Both take the stand that the functioning of all life forms is goal oriented. It is the strictly individual pattern of functioning that is, for the biological theory, the foothold of all intrinsic and organic developments. For the view that sees the ontopoiesis of life at the fore, it is also the intrinsic patterns of operations, moves, processes that builds the organism and carries the dynamic stream of life within its individual and its species' frontiers, and beyond in mutations.

Given this basic agreement, the biological theory of Boorse offers to the phenomenology of life a biologico-physiological substantiation of the concrete hierarchical organization of the functions of the living being, while the phenomenology of life's concept of the ontopoiesis of life gives to an otherwise static or stationary conception of that functional organization the dynamics of the unfolding of life.

The crucially significant coincidence of intuitions lies, however, in the principle of the "norm" of health and disease. In fact, Boorse's idea of species design, which involves an internal, goal-oriented concatenation of functions, meets squarely with the idea that I have been voicing for a long time of the "ontopoietic design" that directs the self-individualizing unfolding of life. While the concept of ontopoietic design offers an approach, through positing a set of transformatory virtualities, to the nature of the moves, operations, processes that carry life, that of species design emphasizes the concreteness and empirical singularity of a life form's basic organization.[31] In these divergencies the two concepts complement one another, providing a twofold pattern of life's progress and as such a norm for its natural course.

While the biological theory of Boorse emphasizes the idealized empirical singularity of the species design for each individual, the phenomenology of life's concept of ontopoietic design emphasizes the play of the intrinsic virtualities with circumambiant forces and circumstances in the living being's accomplishment of its aims. This last applies to the singular route of self-individualizing life, not in the empirical sphere, but in its innermost sphere of virtual reference. It is the unique status of the concept of ontopoietic design that accounts for the dynamic and yet rational nature of beingness-in-becoming. Indeed, it is not without some trace of classical ontology, the ontopoietic, onto-self-forming sphere.

What seizes our attention is that species design is strictly correlated with a reference class of beings. Boorse see this reference class as the fruit of statistical approximation. Species design, in fact, defines its reference class; it is arrived at through an "idealising statistical process." This is as much as to say that the species design principle is – despite the strong emphasis put on its empirical nature – an "idealized" notion.

This statistical idealization calls for some investigation. It is in fact quite enigmatic as to how we come to universalize on the basis of some features of a certain number of cases, leaving to the side those that do not share them, and in virtue of what we form a cogent, structured pattern that applies to all members of a class. Is this universal pattern arrived at through an intentional apprehension of the mind? In that case, its organizing would reflect the structuring patterns of intentionality as the main apparatus of the human mind. Or is it the fruit of a Platonic or early Husserlian "eidetic" intuition which forms – or grasps – an "ideal" pattern of things and being, an eidos intrinsic to their being? Well, in the first case, we would rely on the human mind's specific intentional system and would not be able to account for the fact that patterning in the practice of medicine corresponds to the concrete, biological, factual pattern of the body in its functioning. And yet this fact is most significant for philosophy and medicine alike. In the second case, that of the apprehension of an eidos, the eidetic insight into the nature of the concrete would yield an ideal structure of a static, unchangeable nature and fitting that to the concrete, factual living body would meet with all the difficulties of the Platonic methexis. And yet what Boorse calls the "idealized patterns of the species design" do fit concrete situations. Furthermore, although Boorse see them as rather static patterns, they can accommodate the changeable, transformatory functioning of the living organism. In conclusion, species design cannot be dispensed with, but it does not reside – as Boorse would have it – uniquely in itself. Congruent with our ontopoeitic design of self-individualizing life, it finds in the latter its due continuation and completion.

Let me at this point succinctly say what I have been developing in my previously cited work: what I understand by the "ontopoietic design of life." Refering to Merleau-Ponty's reflection upon the first – and last – principles of the origination of forms – or rationality – within the world of life (*Visible et Invisible*) in which, dismantling the role of essences and ideas, he seems to acknowledge an "intentional arch" subtending the entire edifice of objectivity, I suggest that there is such an arch within the universal schema of the unfolding of life, although of a different nature from Merleau-Ponty's interpretation. Life is a principle of all order. It operates within each unfolding beingness as its formative, and operational intrinsic design which being the primogenital initial endowment of the emerging beingness outlines the set of its virtualities for unfolding; virtualities, vibrant with life, spontaneities, flexible and adaptive propensities and constructive devices. Each and every emerging being does it on account of its specific and unique design in which simultaneously the type and the individual are projected. The self-individualising progress of the emerging beingness stems from this initial "package"; it simultaneously accounts for the progress of the emerging beingness as a "being," as an "entity," and for its self-same living individuality.

The onto-poietic design accounts for the "making" of the living beingness as much as for its formation within this making. It is not disjoined from others; all formations hold together in the entire system of life with its relevances in cosmos.

Indeed, the ontopoietic design is arrived at neither through intentional grasp nor by the idealization of an eidetic insight, but by way of an intuition that aims directly at the constructive unfolding of individualizing life. This intuition, albeit a human act stemming directly from the human creative impetus, the creative act, in a unique way participates in the constructive unfolding of an individual living being even as it is the fulcrum of the human existential condition within the realm of life itself. I propose that it is the design of the self-making of life, of its virtualities for unfolding into different types of living beings that is the natural extension of Boorse's concept of species design.

Indeed, does not the extraordinary fact that the "idealized" pattern of the species design coincides with empirical situations such that it allows an adequate physical intervention in the organism's functioning point to a rational, dynamic networking through the whole of physical, physiological, biological reality as well to the constructivism of the human mind? The ontopoietic design is precisely such a system of reference for life's unfolding, given its positing a universal set of virtualities intrinsic to the individualizing living beings which are yet to be differentiated into types – species – of living beings. It adds to our explanatory power, spelling out what species design implies.

Both the biological theory and the phenomenology of life aim at the operations, processes, workings of unfolding life. The concept of species design focuses on the constancy of operational schemas, on the norm. So does that of ontopoietic design; however, here this constancy is seen as being in progress, in the flux of the dynamic course in which the concrete living being, although grounded in intrinsic virtualities, is transformed as it adjusts to environmental forces and follows its own impetus.

While the focus on species design aims at grasping constant states amid this progress, a focus on ontopoietic design aims at projecting the dynamics of the unfolding, constructing-in-progress course.[32] While the functional system of organic species design is the ground on which this unfolding and game of the fates is played out, to sketch the ontopoietic design is to present the developmental phases of the living beings' individualization-in-progress. Species design has its relative constancy, which remains in place, but the living individual which it grounds "becomes" in unfolding, the self-unfolding of its virtualities. Since the emphasis of our conception of the ontopoietic design falls on self-individualization, species design plays an essential role in the performance of this selfhood, without, however, encompassing it. As a constant (with some evolutionary variations naturally over the span of time), species design serves as a reference system only to concrete, empirical organic functioning. The "telos" of a living being is pursued in its ontic swing rather than in the particular operations that sustain it. Self-individualization is something deeper in the constitution of life, of being as such. It embraces the inner virtualities of the individual becoming of life and the reasons specific to its unfolding in adjusting to changing conditions. In bringing both conceptions together we would have a complete essential system of reference for the processes of life's construction.

Could we, without such an overarching ontopoietic approach, strike the inner chord of the distinction between health and disease? Is this not of paramount importance for the establishment of the notion of "normalcy"? These questions are directed at the fact that medical taxonomy refers essentially to the differentiation of stages in the development of the living individual and to laws of growth, procreation, decline that see developmental transformations in the organs themselves. The dynamic, transformatory course of individual life must be granted an essential role in the defining of normalcy.

To complete our discussion, we enter into the great issue that emerges here at the encounter between species design and the ontopoietic design of life. This is the issue of "human nature." On the one side, Boorse, as stated above, considers the functionally defined species design to represent human nature.[33] But he does not give any allowance for "humanity" and how it constrasts with other species of living beings. Species design can be considered, as a matter of fact, to represent all types of living individuals. On the other side, in the phenomenology of life, the very possibility of intellectually reaching the ontopoietic design lies in the recognition of the specifically human type of living being, one whose radical distinction is the human creative act.[34] Here is recognition of the human creative condition, as emerging from and unfurling a specific set of virtualities with the evolution of life, within nature-life in its progress. In the creative act of the human being, the inner workings of nature meet the creative orchestrations of the human mind, an encounter that makes science and medicine possible.

It is by following the meanders of the creative self-individualizing progress of the human life course that the ontopoietic design comes into focus. Creative virtualities differentiate the specifically human being from the host of other living beings our senses detect. With the human mind and its faculties we have a novel and unprecedented significance of life, one that is uniquely human.

The ontopoietic design of the individualizing mind therefore stretches far beyond biologico-organic laws and constructivism. It naturally embraces the entire spread of the significant transformations that the creative individual projects in forming its very own meaningful universe of life. The avowed aim of Boorse's reducing everything to biology, physiology, the mind, to the brain, has been to show that medicine, in order to determine the "normalcy" of the condition of health, need not consider anything besides the internal functioning of the organism as it faces its environment in order to establish what is the norm. This amounts to eliminating from the conception of normalcy all factors of evaluation – aesthetic, moral, cultural values, lifestyles, tastes. All medical issues, then, are to be framed on the level of the organic. The hope is that by eliminating all "subjective" or "personal" bias, we would concentrate on a strictly "objective" state of affairs.

Can human nature be limited to organic functioning? When it comes to human nature, species design is but the ground of our considerations. From that ground we may descend into the inner workings of nature-life, or we may go beyond to the specifically human constructivism of social, cultural, and personal existence. In short, although at the basic organic level we may and do find constant developmental laws and lines, yet beyond that there spreads, like a peacock tail, the vast spectrum of the specifically human: the aesthetic, intellectual, moral significance of life, which plays a great role in establishing a notion of normalcy or "human nature" as Boorse understands it. With this we enter into the criticism of Boorse offered by Lennart Nordenfelt: If human nature is the norm for health and disease, can this transcendent sphere of human existence be ignored? Will not medicine itself reinstate it?

We then turn to Nordenfelt's holding up of human action as the norm for defining health and illness.

PART TWO
THE TRANSCENDENTAL PLATFORM OF HUMAN VALUES AND CULTURE

1. *Nordenfelt's Standard of Human Action as a Representative Perspective*

In radical contrast to the point of view that Boorse represents, Lennart Nordenfelt proposes making human action the reference point for considering what constitutes health and disease.[35]

Nordenfelt makes this proposal not only to avoid the unwarranted exclusivity of the biological model of health, but also to go beyond common denominators and take in all of the ways, however peculiar, in which human health and illness manifest themselves. He emphasizes the specifically human ways of interpreting life and leaves to the side the significance of the organic level of the living being as such. He establishes action as an axis on which all human issues turn.

In principle there is a certain parallelism between these two theories. Nordenfelt opposes Boorse's conception of health and disease for focusing solely on the internal states of the organism and ignoring "the performance of the organism in a larger, for instance, social context."[36] Basic to Nordenfelt's outlook, then, is human action operating in a social context. Health is defined basically as the ability to act, the ability to function normally.[37] That is the standard for determining adequate internal organic functioning.

We cannot here enter into a detailed analysis of Nordenfelt's theory, in which many distinctions are made. What counts in this discussion is the focus on human action and the focus on organic functioning, there are parallel emphases on goals. Nordenfelt asks, "What is the set of goals, and by whom are they set, which define the abilities that constitute health?" His answer is that these are "the vital goals of man."[38] Although "the vital goals of man can be deduced from his basic needs," they "are identical with the goals he sets for himself during the course of his life."[39] While the first part of the definition of vital goals calls for a "theory of basic needs," the second requires a "subject-goal theory."[40] Ultimately, Nordenfelt defines vital goals as "goals whose fulfillment is necessary and jointly sufficient for the *minimal happiness* of their bearer."[41] He speaks here of "the welfare theory of health."[42]

Above all, this theory is meant to be "holistic." As Nordenfelt writes, the concept of health and disease refers to the human being and not merely to its organism, a "human being as a socially integrated agent who performs a great number of daily activities and is involved in many personal and institutional relations."[43] Hence the norm for health and disease is sought not in organic functioning, but in a psycho-sociological perspective. Nordenfelt enlists in support of this approach Galen, Canguillem, and even the distinguished sociologist Talcott Parsons, quoting the latter as writing, "Health may be defined as the state of optimal *capacity* of an individual for the effective performance of the roles and tasks for which he has been socialized."[44] The feelings of pain and suffering as subjective phenomena, are related to the subjective experience of well-being as much as to objective ability (or disability) to perform a task.

We may say here that due attention is given to the suffering that belongs to disease, that the focus here is on health and illness rather than on disease as such.[45]

It is through the notions of action and ability that Nordenfelt expects to gather together all the psychosocial functions of the human being that impinge on the notions of health and illness and may help us understand the state of affairs. Action, understood as being goal-oriented, is but a segment of the agent's "plan to reach certain goals."[46] Hope of fulfilling a plan generates an interlocking series of purposefully stratified actions. Each of them belongs to a network, "while there is a basic action that initiates the chain."[47] Nordenfelt discusses the path from basic action to the "end state" of the plan's fulfillment in relation to the situation in the agent's environment: "Very much of what we do that entails actions beyond basic action is dependent upon the course of the world outside ourselves." He distinguishes "standard" and "extraordinary" circumstances to which dis-ability would be related in the natural and cultural environments, saying that their definition depends upon "society's profile of goals."[48]

The vast ramifications of these social and psychological interdependencies lead Nordenfelt to consider the concept of health in a Wittgensteinian fashion as a "family concept" within which there are variations, with each variation having a resemblance to all the others.

Vital goals corresponding to basic human needs serve to clarify the concept of health. We have to emphasize that in speaking of the necessary vital needs we cannot remain at a biological level. "Standard circumstances" that stem from environmental and cultural situations have to be considered necessary conditions for survival, therefore. In our contemporary Western world, for example, the requisite to satisfying basic physiological needs is a means for obtaining funds for the purchase of food, shelter, clothing, which involves an occupation. The concept of "basic need" then stretches into the social world. However, and this a point of paramount significance for our discussion, basic goals are not to be thought of as being engrained in individuals' natures or imposed on them, but are proposed by individuals, personally. Basic goals are thus an expression of the person's innermost nature, inclination, taste, values, etc.

In his attempt at framing a theory encompassing the issues of the human being's survival in the physical and societal world, Nordenfelt's basic analysis of action and its stratifications arrives at what he calls "a welfare notion of health." First of all, health is a capacity to pursue in action one's personal goals through to their fulfillment. Nordenfelt defines the welfare concept of health in the following fashion: "A is healthy if, and only if, A is able, given standard circumstances in his environment, to fulfill those goals which are necessary and jointly sufficient for his minimal happiness."[49] This amounts to assuming "minimal happiness" to be the apex of a scalar network of actions fulfilling basic needs, and, as a matter of fact, to be the basic need itself – the vital need? – and its accomplishment.

Nordenfelt discusses both ancient "objective" conceptions of happiness, whether defined in terms of internal goods (virtues, powers of mind, etc.) as in Plato and Aristotle or in terms of external goods (family, friends, possessions), and "subjective" conceptions of happiness that, following Tatarkiewicz, define it in terms of "gratification." He quotes Tatarkiewicz,

"A life is a happy one if we are satisfied with it, and it is the man satisfied with his life who is qualified as happy. Whether he possesses goods and what kinds they are makes no difference. As long as he is content, he is happy."[50] But this "gratification" or "contentment" calls for further qualification. The object of the feeling of contentment or happiness should be the person's existence as a whole. Furthermore, happiness has quantitative and qualitative dimensions of duration and richness.

With this in view, the vital goals of man, which are at the heart of this inquiry, should be judged according to two standards: they should be absolutely necessary for the achievement of happiness, and highly probable, to "contribute to everyone's happiness."[51] Here we enter into the evaluation of proper and improper goals (improper goals being counterproductive goals, trivial goals, those having a compulsion as a goad, etc.), and the question then emerges of how we can evaluate and with what as a point of reference. What are the conditions of health conceived of in terms of "ability" to accomplish the vital goal of minimal happiness?

2. *The Affinities between the Conception of Health as Ability and the Phenomenology of Life: Human Self-Determination in Existence*

Nordenfelt deliberately intends to frame the question of health in terms of philosophical anthropology. His basic tenet is that the human being is to be conceived as an agent immersed in interaction with others within the world of life. This squares with the pivotal ideas of the philosophy/phenomenology of life. The goal orientation Nordenfelt speaks of accords readily with that philosophy's conception of the ontopoietic design of the specifically human individualization of life.

Nordenfelt's views do appear to be restricted by the canons of description set by analytic philosophy. His conception of health as the ability to achieve minimal happiness, while having implications for action within the human world and calling for the evaluation of innumerable factors and circumstances, falls short of enlisting the entire deliberative "mechanism," which hampers the setting of proper goals. Most critically, it sets to the side the question of triggering the will to pursue them. We need here an "apparatus" that indicates with reference to what and in virtue of what these vital goals – vital needs going beyond the organically determined needs of survival – are to be formulated, individuated, and actively pursued.

I will now succinctly propose the main points at which Nordenfelt's theory and the phenomenology of life coincide directly and those where

Nordenfelt's theory calls for completion by the phenomenology of life's concept of the overarching ontopoietic design of life and the points of reference therein provided.

In contrast to the reductionist biological platform for a theory of health that finds the norm for health and disease in a species design that is but the intrinsic pattern of the organism's typical internal functioning and leaves out all other perspectives on human life, the theory of action extends over the entire functioning of the human being, including the psychological and psychosocial "historical" spheres that make human life specifically human.

Although the perspectives and approaches of the respective theories are radically different, nevertheless I see them as being complementary contributions to a fuller conception of "human nature," one that embraces the developmental phases of the evolution of life and the transcending specifically "human creative condition." I have tried above to show how the main tenets of the biological conception of health coincide with those of the ontopoietic conception of life, and now I will attempt to show in a succinct fashion what affinities the action theory of health maintains with the specifically human self-individualization, self-interpretation of life. The specifically human phase, albeit grounded in the biological sphere of vital functioning, extends into what I have called "man's self-interpretation-in-existence."[52] Extending over both of these spheres, the one "vital" and the other "creative," and all between, there extends the ontopoietic design of life.

This conception of life in its evolutive differentiation into types/species offers an overarching reference system for both. First of all, the theory of action sees the human being as engaged in an active, self-devised and self-enacted pursuit of happiness. Here, the emphasis on the self-initiating pursuits through series of actions that intergenerate as they advance toward the fulfillment of a goal coincides remarkably with the ontopoietic conception of the self-individualization of life with its emphasis on the selfhood that maintains itself through innumerable obstacles in the forces of the environment or submits to them and changes course. Yet the description of action given in this theory centers on the *behavioral* aspect of life pursuits in a fashion close to the one analytic philosophy uses to describe reality, and does not ask after causes and sources, and this is to observe the mere mechanism of acting.

Although Nordenfelt emphasizes the hierarchical, generative organization of the actions that lead toward a goal, nevertheless his conception needs certain questions and answers. Why and in virtue of what is a human being goal-oriented? And we must ask, if we are not to fall back on survival and procreation-oriented norms, what is the sense, the significance of goals that the person sets for him/herself? In what way do they affect and penetrate human life, so that the agent may experience contentment or happiness? To grasp human nature we simply cannot remain at the level of comportment. In virtue of what are living beings "agents" capable of deliberating and setting goals? Indeed, in virtue of what do our goals correspond to our innermost desires, aspirations, visions? All this amounts to asking the great question of what is the continuing sense that emerges with purposeful action, such that we may speak of our *own* life as a whole and of our existence being either fulfilled or a failure. It is in view of this continuity of a human person's life, which is not to be identified with the empirical course of life, that life becomes a project.

It is this full-fledged sense forging and sense acquiring function of human action that is comprised by the concept of "human self-interpretation-in-existence." Here is described the specifically human phase of self-individualizing life, the specifically human phase of life's ontopoietic design, which however, and here falls the emphasis, is self-devised and proposed by the agent.[53]

As the ontopoietic design carries on the biological, vital, transformatory dynamisms of living progress, it passes from organic vitally significant complexities to the originary endowment of the human being, creative virtualities that burst forth with imaginative impulses, sensations, ideas, initiatives, the human mind and the freedom it delights in, the meanings it finds.

Only when we envisage this creative unfurling and enter into activities that are creatively significant, may we understand and account for the continuity of pursuit that makes life the experience of a whole. Indeed, to be able to consider, to experience, to "see" and evaluate life as a whole as being happy or not, one must be able to see a continuity within the more or less loosely connected fragments forming the sequences of active comportment and within singular goals. Like Beckett's "Man at the Sea," we often cannot distill from the concrete segments of our project-pursuit a deeper sense that they have for us and which gives a unifying meaning to the whole "itinerary." If we did not obtain this sense from a new perspective, we would see ourselves as being futilely involved in a series of dispersed events prompted only by a neutral and senseless play of forces, like Beckett's "Man at the Sea."

Thus, Nordenfelt's theory of action calls out for and even points to an innermost core of sense-bestowing, helping agents crystallize and express the meaning of the activities they are engaged in. The person thereby crystallizes and expresses his or her individual tendencies, aspirations, talents, taste while unfolding a personality. This inner core, as a virtual factor of sense, is shaped, advanced or stunted, raised or dwarfed, induced to blossom or paralyzed in the encounter between the forces of the circumambiant world and inner strengths and weaknesses. When we speak of the "whole" of life as being the accomplishment of happiness, even minimal happiness, it is the continuity in this inner forging of ourselves that we refer to.

In other human beings, creativity endows their lives with meaning beyond survival and procreation through the enactment of their stories.

For the last two decades I have been advancing this understanding of "human nature." This conception of specifically human nature as being of a uniquely "historical" nature, being actualized from a set of virtualities and projected from the innermost core of human creative functioning in specifically human "self-interpretation-in-existence," brings out the continuity in the inner core of meaningfulness. No matter how many times we change our projects, we maintain an innermost continuity of meaningful pursuit, the continuity of our individual life, of our personal unfolding. When we change one of our basic projects, we say we are "starting life all over again," which means that this new turn affects our entire sense of the unfolding of our life and marks another of its phases.

Historicity, which distinguishes the human being from all the other living beings our explorations have found and is the innermost thread of continuity binding all our individual, personal significant moves together, *follows life's ontopoietic design and simultaneously creatively crystallizes it in the enactment of our self-unfolding*. As noted above in our discussion of the thinking of Boorse, it constitutes the specifically human ontopoietic self-individualization in self-interpretation-in-existence.

But above all it is the human creative mind, our creative condition, that introduces human significance into the realm of nature, drawing on the specifically human virtualities of perception, deliberation, evaluation, judgement, imagination.

CONCLUSION
THE ENTENTE BETWEEN MEDICINE AND PHENOMENOLOGY OF LIFE AND ITS ONTOPOIETIC ARCH

The seemingly radically opposed theories of health under discussion present two norms by which medical diagnosis and therapy may be governed. However, ambition has carried both of these theories beyond their reach, for neither comprehensively treats human nature.

In practice, these diverging approaches supplement each other. The approach emphasizing biological functioning in concrete cases needs assessment of the social situation of the patient. And the approach that focuses on social action cannot proceed without probing the organic endowment wherein the impediments to purposeful acting are to be found.

However, the thrust of our argument has been that now both of these approaches in their incompleteness point to the indispensable reference system of the philosophy/phenomenology of life, which brings all of the strings of individualizing life together.

To summarize my argument, let me say that organic, physiological functioning points to an ontopoietic design stretching over all of the self-individualization in nature-life. In this context, the biological functioning that Boorse concentrates on points to the laws of individualization at work in life and the action framework for conceiving of the human being constructed by Nordenfelt points directly to human self-interpretation-in-existence, to the sense-originating function of the human creative mind.

I hope that I have succinctly shown that these outstanding and representative theories guiding medical discussion of health and disease each coincide in significant ways with the phenomenology of life's apprehension of its ontopoiesis, and further that these two diverging approaches are bridged in the insight that life's ontopoietic design subtends, suffuses, and governs the separate realms of human existence that each takes as its base.

The norm for defining health and disease should be sought in the overarching reality that holds all the strings unfolded in a human lifespan.

The World Phenomenological Insitute

NOTES

[1] Anna-Teresa Tymieniecka, *Logos and Life*, Book 1: *Creative Experience and the Critique of Reason*, Analecta Husserliana, XXIV. Dordrecht: Kluwer Academic Publishers, 1988.

[2] See Henk A. M. J. Ten Have, *The Growth of Medical Knowledge*, Philosophy and Medicine 36. Dordrecht: Kluwer Academic Publishers, 1990.

[3] Anna-Teresa Tymieniecka, "Nature-Life and Culture in the Unity-of-Everything-There-Is-Alive," in Marcelo Sanchez Sorondo and Dario Antiseri (eds.), *Physica, cosmologia, naturphilosophie: nuovi approcci*, Collana Dialogo di Filosofia 10. Rome: Herder, Università Lateranense, 1993, pp. 292–298.

[4] For the moral disarray, see Anna-Teresa Tymieniecka, "The Golden Measure," in Anna-Teresa Tymieniecka (ed.), *Life: The Ideal of Human Kind*, Analecta Husserliana XLIX. Dordrecht: Kluwer Academic Publishers, 1996, pp. 3–25; see also Tristram Engelhardt, "From

Abortion to Euthanasia: Rethinking the Value of Life at the End of the Christian Age," in the present volume.

[5] For a brief history of medicine, see Guenter B. Risse, "History of Western Medicine from Hippocrates to Germ Theory," in Henry Ernest Sigerist, *On The History of Medicine*, New York, N.Y.: M.D. Publications, 1960.

[6] Rudolf Virchow, "One Hundred Years of General Pathology," in his *Disease, Life and Man*, trans. Lelland J. Rather. Palo Alto: Stanford University Press, 1958, pp. 170–215.

[7] Rudolf Virchow, "Cellular Pathology," in *ibid.*, p. 85.

[8] Rudolf Virchow, "Atoms and Individuals," in *ibid.*, p. 133.

[9] Anna-Teresa Tymieniecka, "The Great Plan of Life," in Anna-Teresa Tymieniecka (ed.), *Phenomenology of Life and the Human Creative Condition*, Book 1: *Laying Down the Cornerstones of the Field*, Analecta Husserliana LII. Dordrecht: Kluwer Academic Publishers, 1998, pp. 3–29.

[10] See Arthur L. Caplan, "The Concepts of Health, Illness and Disease," in his *Concepts of Health and Disease*, Reading, Mass.: Addison-Wesley, 1981, pp. 234–247.

[11] Christopher Boorse, "Health as a Theoretical Concept," *Philosophy of Science* (1977) 44: 4 511–573.

[12] See Zbigniew Szawarski, "Two Models of Medical Knowledge," in the present volume.

[13] See Anna-Teresa Tymieniecka, "Ontopoiesis of Life as the New Universal Paradigm; An Alliance between Phenomenology/Philosophy of Life and the Sciences of Life," "Theme" in Anna-Teresa Tymieniecka (ed.), *Life-Scientific Philosophy*: *Phenomenology of Life and the Sciences of Life*, Analecta Husserliana LIX and LX. Dordrecht: Kluwer Academic Publishers, 1998, 1999.

[14] See Boorse, *op. cit.*

[15] See Lennart Nordenfelt, *On the Nature of Health, an Action-Theoretic Approach.* Dordrecht: D. Reidel (Kluwer), 1987.

[16] Boorse, *op. cit.*, p. 554.

[17] O. Temkin, "Health and Disease," in *Dictionary of the History of Ideas*, Vol. 2, New York: Scribner's 1973, pp. 395–407; quoted in Boorse, *op. cit.*

[18] Boorse, *op. cit.*, p. 494.

[19] It appears that there is a universal consensus on the usefulness of statistics in medical diagnosis. Normality is understood then in terms of statistical normality and variation. However, Boorse (*op. cit.*, pp. 546–47) and Edmond A. Murphy, while recognizing the indispensability of statistics, are troubled by the vagueness of statistical normality, for which Murphy blames "epistemic, ontological and under-conceptualization difficulties" (in his *The Logic of Medicine*, 2nd ed. Baltimore: The Johns Hopkins University Press, 1997, p. 155).

[20] Boorse, *op. cit.*, pp. 556–558.

[21] *Ibid.*, p. 556.

[22] *Ibid.*, p. 555.

[23] *Ibid.*, pp. 555–558.

[24] *Ibid.*, p. 555.

[25] *Ibid.*, p. 558.

[26] *Ibid.*, p. 560.

[27] *Ibid.*, p. 557.

[28] *Ibid.*,

[29] *Ibid.*

[30] *Ibid.*, pp. 561–62.

[31] See Tymieniecka, "Ontopoiesis...," *op. cit.*

[32] Tymieniecka, "Nature-Life...," *op. cit.*, pp. 292–299.

[33] Boorse, *op. cit.*, p. 563.

[34] See Anna-Teresa Tymieniecka, "The First Principles of the Phenomenology of Life: Charting the Human Condition," in Anna-Teresa Tymieniecka (ed.), *Phenomenology and Philosophy of Life (The Spanish Perspective)*, Analecta Husserliana XXIX. Dordrecht: Kluwer Academic Publishers, 1986; and Tymieniecka, *Logos and Life*, Book 1, *op. cit.*

[35] At the very outset of our presentation and interpretation of the contrast between these theories, it must be noted that at the bottom of it is a distinction made by both of these authors, albeit in a different fashion, between the concepts of "disease" and "illness." The ways in which they emphasize one or the other are essential to their theories. We cannot here enter into a discussion of these concepts in depth and at length. Let us assume that, in general, by "disease" is to be understood a defective condition within the organism and by "illness" the manifestations of that defectiveness in the malfunctioning of the organism.

This distinction is viewed differently in the theories of Boorse and Nordenfelt. We cannot enter here into the intricacies of a distinction to which Boorse has devoted a special study, "On the Distinction between Disease and Illness," *Philosophy and Public Affairs*, Vol. 5, I, 1975, pp. 48–68.

In general, for Boorse disease is a basic defect in the organism that might or might not manifest itself directly as disfunction in the organism ("On the Distinction...," pp. 50–56). Since he emphasizes that the criterion for health is the functioning of the organism and its parts in accordance with species design, it is disease – whether manifest or not – that is his focus. It is likewise in functional terms that he addresses abnormal mental conditions, integrating them into his theory by positing "natural mental functions" and holding that "recognized types of psychopathology are unnatural interferences with these functions" ("On the Distinction...," pp. 56–60). (This stands in contrast to the opinions of psychiatrists such as Thomas Szasz who see no criteria by which mental conditions popularly considered pathological can be considered "mental diseases.")

And while Boorse focuses on "disease," Nordenfelt, who stresses primarily the capacity of the human being for action, favors the concept of "illness" when speaking of disfunctions.

[36] Nordenfelt, *On the Nature of Health, an Action-Theoretic Approach, op. cit.*, p. xiii.

[37] *Ibid.*

[38] *Ibid.*, p. xv.

[39] *Ibid.*

[40] *Ibid.*

[41] *Ibid.*

[42] *Ibid.*, pp. 77–81.

[43] *Ibid.*, p. 35.

[44] *Ibid.*, p. 97.

[45] *Ibid.*, p. 36.

[46] *Ibid.*, p. 37.

[47] *Ibid.*

[48] *Ibid.*, p. 49.

[49] *Ibid.*, pp. 79.

[50] *Ibid.*, p. 81–82.

[51] *Ibid.*, p. 91.

[52] For my conception of human self-interpretation-in-existence, see my *Logos and Life*, Book 1: *Creative Experience and the Critique of Reason*, Analecta Husserliana, XXIV. Dordrecht: Kluwer Academic Publishers, 1988; and Book 3: *The Passions of the Soul and the Elements in the Ontopoiesis of Culture. The Life Significance of Literature*. Dordrecht: Kluwer Academic Publishers, 1990; but specifically my shorter study, "Life's Primogenital Timing," in Anna-Teresa Tymieniecka (ed.) *Life: Phenomenology of Life as the Starting Point of Philosophy*, Analecta Husserliana L. Dordrecht: Kluwer Academic Publishers, 1997, pp. 3–22.

[53] For thoughts on accomplishment through self-interpretation-in-existence, see the same works as in the previous note and my studies throughout the Analecta Husserliana series, beginning with my study in Anna-Teresa Tymieniecka (ed.), *The Self and the Other. The Irreducible Element in Man*, Analecta Husserliana VI. Dordrecht: D. Reidel, 1978.

PART I

THE LOGIC OF MEDICINE

ZBIGNIEW SZAWARSKI

TWO MODELS OF MEDICAL KNOWLEDGE

When a physician talks about disease or impaired capacity, he usually means some imperfection in biological and/or psychological functioning of the human being. As a medical student, he was taught some basic knowledge concerning human biology, anatomy, physiology, and pathology. Theoretically he should easily be able to tell the difference between the perfect and the imperfect, the normal and the pathological, the healthy and the sick. A pediatrician sees at once the difference between the healthy and the severely handicapped newborn. A GP can easily diagnose herpes or trivial flu, and has no difficulty in assessing the level of deficiency in our sight or hearing. Of course there are some difficult cases where it is impossible to make even a tentative diagnosis, and when despite all technological support, the physician is unable to find any convincing explanation of the patient's condition [Cassell, 1978]. These are, however, incidental and rare cases and indeed very few doctors will straightforwardly declare their ignorance. By and large medicine, as a powerful branch of modern science, shares its epistemological optimism: even if at present it cannot explain nor treat some illnesses, it is only a question of time before it will be able to.

I do not share this optimism. Of course, it is impossible to deny that in the course of the development of medical science, our knowledge of man's structure and functioning will be increasing. However, I do not think it is possible to reach **a consensus** on what constitutes the fundamental or significant imperfection in the biological and psychological functioning of a human being. And how should this imperfection be treated? It could seem that medical science provides us with a distinct norm that allows us to distinguish between health and illness, ability and disability. Yet this is not the case, because there are various models of medical knowledge. Therefore, our understanding of disease and impaired capacity is determined by the model of medical knowledge that we accept. Consequently, a belief that there exists only one definite model of disease and impaired capacity is an illusion. Our concept of impaired capacity depends directly on the medical philosophy that we are professing.

Nobody questions the existence of various models of medical knowledge. There is, however, no universally recognized terminology for each particular model. As a rule, the scientific model of medicine is juxtaposed with the humanistic model. The mechanist or reductionist model of health is

A.-T. Tymieniecka and Z. Zalewski (eds.), Analecta Husserliana LXIV, 39–47.

contrasted with the system approach or bio-psycho-social model of health, and some authors distinguish between biostatic and holistic models. There is even a recent attempt at a synthesis of both approaches as the infomedical model [Foss]. Since a choice of a suitable convention is, as a matter of fact, entirely arbitrary. For the purpose of this article, I shall use three terms which, in my opinion, aptly reflect the characteristic features of the respective models. I shall thus subsequently present, in a nutshell, the biomedical model, whose paradigm is the biological knowledge of man and the axiomedical (from the Greek *axio*: worth, value) knowledge of the human person and whose characteristic feature is the analysis of the basic medical categories in value terms. I shall not be concerned with the infomedical model, whose main stress lies upon the concept of self-organization and the processing and transmission of information. It is clear that all these models assume a particular philosophy of nature, a definite concept of man [Wulff, 1994], a definite concept of health and disease, and a distinctive idea of examination, diagnosis, therapy, and prognosis.

THE BIOMEDICAL MODEL

It is not my task here to present the history and significance of the biomedical model in contemporary medicine. I only want to point out those characteristics of the biomedical model that are directly relevant to the concept, definition, and consideration of disease and impaired capacity:

1. The model assumes a materialist and determinist model of the world. Life is a form of existence of matter, and all life processes are subject to definite laws. The task of science is to explain and describe each life process, and to formulate relevant laws. This is the subject of biology. Since it is possible to describe each biological system in the language of chemistry and physics, it does not really matter which system is the subject of scientific investigation, for it is assumed that scientific laws are universal and equally relevant to all life processes. As a biological organism, man differs from other living organisms only in the degree of complexity.

2. Individual survival and reproduction are the fundamental purpose for every living organism. Hence the structure and all functions of each organism are subordinate to this general purpose. In this respect, a biological organism does not essentially differ from a machine. Just as the proper function of a clock is to measure time, and the function of a homeostat is to regulate temperature, so the proper function of all living organisms is to survive and reproduce. A broken or unwound clock does not carry out its proper function.

Similarly, a severely damaged living organism is not able to perform its proper function. The understanding of a living organism's functioning is possible through dismantling a complex biomechanism into simpler parts and subsystems which can be examined in detail. Just as each organism is the sum of organs and functions which constitute a complex whole, so medical knowledge is a sum of disciplines which are concerned with particular organs or organic systems of man.

3. The mind and mental life are not necessary conditions for the biomechanism to function properly. "From the very perfection of animal actions", wrote Descartes, "We suspect that they do not have free will" (Rosenfeld, 31). Neither a clock, nor a homeostat, nor any animal possesses a soul. And if man differs from the animal in that he thinks or possesses a soul, then the mind either exists independently of the body (the Cartesian cogito) and it is thus difficult to understand how it can be the cause of physical processes, or the mind is entirely reducible to the material processes of the brain.

4. A human being is healthy if he is able to properly function as a biological organism. As disease means a substantial disturbance in the normal functioning of the organism, a concept of norms thus plays the basic role in the biomedical model. If we assume that it is possible to define a proper norm of the functioning of the human [cf. C. Boorse], this has a fundamental influence on the concept of medical knowledge. Because if we believe that medical knowledge is, as a matter of fact, the sum of information necessary to treat and prevent diseases, then in the biomedical model this knowledge concerns mainly the perfect (proper) and imperfect (improper) functioning of the human body, and physiology and pathophysiology are the theoretical foundations of the biomedical model.

5. Such a concept of medical knowledge directly determines medical practice. Examination thus consists in determining all irregularities and deviations from the norm. Diagnosis concentrates on defining the cause of the irregularities, and treatment focuses on the search for means to restore the organism's biological abilities.

The philosophy specific for the biomedical model is simple to express. The subject of medical activity is a damaged or malfunctioning biological mechanism – the human body [Leder]. The cause of damage may be internal (a virus, bacteria, a genetic error) or external (acquired injury). However, in each case the objective of therapy is to restore the damaged biological function or to replace the lost one. In other words, the main goal of medicine is to preserve life and help regain full ability for survival and reproduction. If, for instance, due to an organic defect, a patient is unable to have offspring in a natural way, they can be easily helped through IVF. If the patient's pancreas does not produce insulin, the patient can nevertheless function normally if he does not neglect regular insulin injections. Similarly, proper diet introduced in time can prevent the onset of such a serious disease as PKN. And if a patient suffers from severe depression, there is not, apparently, any more effective way of treating this than by raising the level of serotonine in the brain through administration of Prozac. It is evident that there are still many diseases (like AIDS, SM, or some cases of cancer) that are not yet curable by modern medicine. It is also true that medicine is, to some extent, also the victim of its own success, due to the effective extension of the patient's life span which results in a growing number of occurrences of old-age diseases. However, the possibilities of scientific progress are theoretically unlimited. And it is not entirely out of the question that Descartes' dream of constructing a living machine will perhaps some day come true. It would be sufficient to be able to decode and reassemble the molecular structure of the ovum and the sperm, and the rest seems dead easy.

What is the connection of the above with impaired capacity? In the biological model capacity is a purely biological category which means simply the organism's ability to survive and reproduce, and that is strongly related to a specific developmental stage. Each more complex living biological system goes through three developmental stages: birth and childhood, maturity and a period of reproductive activity, and gradual decline. From the evolutionary point of view it is essential for the individual to achieve maturity and leave behind the progeny capable of independent existence. Nature does not care about prolonging an individual life of a specimen; what is really important is to preserve the genetic information specific to a species, and to pass it on to the species' next generation.

An individual can be less or better adapted to the environment and, as Darwin observed a long time ago, only the individuals and species that are best adapted have the greatest chances of survival. It is true that humans differ from animals also in that they change their environment in order to meet their proliferating human needs. But exactly this essentially human activity demands not only upright posture and a prehensile hand, but also, first and foremost, a highly organized brain and the ability of critical and creative thinking. We should remember that, in the biomedical model, thinking is a biological category and is impossible without a fully developed and well functioning brain. Severe and irreversible damage to the cerebral system means total disintegration of the entire biological system. It does not mean, however, that it is the properly functioning brain that constitutes the necessary condition for the biological existence of man. Persistent vegetative states are also a form of biological existence, but this kind of existence is only possible when it depends upon active help from other fellow beings. Left on his own, a PVS patient is condemned to certain death. But exactly the same end awaits everyone who, left alone and deprived of the protective cocoon of human culture and civilization, breaks his femur, is bitten by a poisonous snake, or exposes himself to heatstroke or hypothermia. Thus, it does not really matter if we talk about disease, injury, disablement, impairment, genetic defect or handicap. All these states of organism are abnormal, and when treated in biological terms, are *all forms of impaired capacity*. And this is their abnormality which makes them a negative value. Even a trivial impairment in appearance may dramatically hinder finding a partner for marriage, and so prevent having much desired offspring.

It is not difficult to point out the weak points of the model. Apart form the standard critique of the reductionist and mechanical model of man (Wulff, 1994), the basic drawback of this model is the belief that there is a perfect norm for every biological structure and function. A properly built protein or enzyme has a perfect structure a priori. H_2O is always H_2O, and NaCl is always NaCl. Yet, when we proceed from simple chemical compounds to complex biological systems, problems arise. It is true that a gene is a gene, and a chromosome is a chromosome, no matter whether we explore the genetics of the sweet pea or the fruit fly [Cassell, 1992]. It turns out, however, that minute changes in the ends of human chromosomes may bring about dramatic results causing deep and irreversible mental handicaps (*New Scientist*, Feb. 11, 1995, p. 18). Perhaps genetic engineering will soon be able to correct even the small imperfections of chromosomes; however, to determine whether a chromosome is abnormal, *we must set a norm*, and this seems highly contentious. It is possible to argue, as Canguilhem does, that disease is as normal and natural a phenomenon as health, and consequently we may say that it is absolutely normal and natural that in the course of the exchange of genetic material between a male and a female, "small rearrangements and deletions" occur in chromosome ends. It is not a technical problem only, because, paradoxically, if there exists any perfect blueprint at all for the human species, then it assumes the individual variety of the genetic material, as it is the most beneficial to the survival of the species. And it is really difficult to say what constitutes the perfect human genotype and what its imperfections would consist in.

Moreover, assuming that disease dwells in the body, and that the body is actually the patient, this model completely ignores the significance of the psychological factor in the etiology and treatment of diseases. It is this element that, linked to the social factor, plays a fundamental role in the axiomedical model.

THE AXIOMEDICAL MODEL

This model assumes the system approach to nature as an extremely complex natural system composed of numerous intertwined subsystems. All these systems have certain common characteristics. As H. Brody says: "(1) Natural systems are wholes with irreducible parts; (2) natural systems maintain themselves in a changing environment; (3) natural systems create themselves in response to challenges of the environment; (4) natural systems are coordinating interfaces in a hierarchical organization" [Brody]. The category of harmonious and dynamic equilibrium is the basic theoretical notion in this model; it characterizes the properly functioning hierarchy of natural systems.

Man is a natural system, and as such constitutes an arrangement of subsystems that belong to descending levels in the hierarchy, including such subsystems as the nervous system and particular organic systems, tissues, cells, up to molecular structures and subatomic particles. On the other hand, however, a human being is a subsystem in a hierarchy of more complex, ascending wholes, such as a relationship of two persons of opposite sex, a family, a community, culture, society, nation and mankind [Laszlo, Brody, Sheldon, Querido]. A human is thus something more than just a body, a biological organism or a sum of biochemical properties. It is a being that feels, thinks, makes decisions, communicates and shares emotions with other fellow beings; in short, a being that creates and functions within *a system of values*. This fact changes, in a fundamental way, the notion of man's attitude toward biological survival. For although it is true that survival and reproduction are the main biological goal of each living organism, the difference between man and animals lies in the fact that *man is a person*, man as a free and rational being individually lends sense to his existence, and it often happens that other values are incomparably more important to him than mere biological survival and reproduction. The goal of human existence is thus not a purely animalistic biological survival, but a survival worthy of the human person. And it is the human individuum that defines what it means for *him* to "survive as a human person".

Health cannot be reduced merely to the proper functioning of only one biological system; it constitutes a state of the dynamic equilibrium of many systems that comprise biological, psychic and social levels of an individual's functioning. This model is often called "a biopsychosocial model of health". I do not think that this term is correct, because the most important feature of this model does not lie in the fact that an individual participates in many interdependent systems, but in the fact that an individual defines and fulfills the goal of his existence as a human person through the choice and fulfillment of particular sub-goals. Therefore I propose to call this model "the axiomedical model".

L. Nordenfelt's equilibrium theory of health is an excellent example of such a model. Virtually, the theory boils down to several simple statements:

a) A person P is completely healthy if, and only if, P is in a bodily and mental state such that he or she has the ability to realize all his or her vital goals, given standard circumstances [Nordenfelt, 1987, pp. 76–80].
b) A person P is unhealthy to some degree if, and only if, P, given standard circumstances cannot realize all his or her vital goals or can only partly realize some of them.
c) X is a vital goal of P's if, and only if, x is a state which is necessary for P's minimal happiness.

[Therefore:]

d) To be healthy is tantamount to having the ability, given standard circumstances, to realize one's minimal happiness [Nordenfelt, 1993].

Still, because the total set of standard circumstances is defined in relation to a certain time and cultural situation, the notion of health is culture-relative, and "one and the same person with the same physical and mental capacities, can in one culture emerge as healthy but in a very different culture as ill" [Nordenfelt, 1993, p. 100].

Stressing the overwhelming significance of value in both the individual and social aspect, the model avoids the dramatic opposition between biologically defined disease and health, and highlights the notions of wellness and illness. A human person feels either well or ill, she is either happy or suffering. Besides, it is important that this subjective sense of one's own predicament may, but does not have to necessarily be, tied to disease, disability, or physical impairment. Disease or severe disability may, but does not have to be, an obstacle to a happy and successful life. And it is also possible that it is disease and illness that give an individual the sense of a creative and successful life [Pickering]. Good examples are the great novel by T. Mann, *Dr. Faustus*, and the life of S. W. Hawking.

If a mechanical fault or breakdown of the human biological system is not the sole cause of illness, but the cause is also conditioned socially and psychologically, and if the sense of being well or healthy virtually depends on the individual's ability to fulfill his vital goals, then this significantly influences the model of medical knowledge. Medicine then ceases to be value-free knowledge, and becomes a programmatically normative discipline [Wartofsky]. It does not discover the norm, but in a way it projects it [Canguilhem]. It bears obvious consequences for medical practice. Examination not only consists in identifying all anatomical and physiological deviations from the norm, but also considers, to a great extent, the patient's narrative. Diagnosis is not confined to merely determining the direct physiological causes of illness, but analyses the broad spectrum of complex psychological and social determinants. Therapy, in turn, stresses the significance of the patient's hidden psychological powers, that is, the so-called psycho-immunological factor. The patient is no longer a passive body subject to medical manipulation, but becomes an autonomous partner in his own right in the medical encounter.

The greatest advantage of this model is, no doubt, the discovery that the patient is something more than a body in pain only, that he has the individual sense of life and his own world of values. On the other hand, recognition that the patient and only the competent patient has an ultimate say in choosing his vital goals reduces the art of medicine to an ordinary craft. The doctor ought to know what to do to make his patient happy, but it is not for him to pass any judgements about his patient's idea of happiness. Consequently, if a 58-year-old woman wants to conceive a child, let us give her a child. If a young woman wants to be the first living example of the third sex, let it be. It is so easy to do hysterectomy and mastectomy. And if someone is tired of life, full of pain and suffering, and wants to die, a good doctor will provide him effective help in dying. In the axiomedical model of medical knowledge, what is good for the patient is more important than the knowledge of anatomy.

CONCLUSION

I have presented two models of medical knowledge. The first model is grounded on the outside knowledge of the patient and is pertinent to scientific evidence or evidence based medicine. It sees the patient mostly as a complex biological system and tends to neglect the spiritual dimensions. The second one features the inside knowledge of the patient and his or her idea of a good life, but its main disadvantage is a sort of epistemological and moral relativism. Perhaps one day we shall reach a synthesis of both approaches in a comprehensive system of medicine practised as art and science. However, this will need a new philosophy of life.

University College of Swansea

REFERENCES

Baron, J. R., "Why Aren't More Doctors Phenomenologist?" in: D. Leder (ed.), *The Body in Medical Thought and Practice*, Kluwer Academic Publishers, 1992.

Boorse, C., "Health as a Theoretical Concept", *Philosophy of Science* (1977) 44, 542–573.

Brody, H., "The System View of Man: Implications for Medicine, Science, and Ethics", *Perspectives in Biology and Medicine*, Autumn, 1973.

Canguilhem, G., *The Normal and the Pathological*, New York: Zone Books, 1991.

Cassell, E., "The Conflict between the Desire to Know and the Need to Care for the Patient", in: S. F. Spicker (ed.), *Organism, Medicine, and Metaphysics*, D. Reidel Publishing House, 1978.

Cassell, E., "The Body of the Future", in D. Leder (ed.), *The Body in Medical Thought and Practice*, Kluwer Academic Publishers, 1992.

Foss, L. "Putting the Mind back into the Body. A Successor Scientific Medical Model", *Theoretical Medicine* (1994) 15: 291–313.

Greaves, D., *The Historical Conceptualization of Impaired Capacity and Some Ethical Implications*. Unpublished manuscript.

Laszlo, E., *The Systems View of the World*, New York: Brazilier, 1972.

Leder, D., "A Tale of Two Bodies: The Cartesian Corpse and the Lived Body", in D. Leder (ed.), *The Body in Medical Thought and Practice*, Kluwer Academic Publishers, 1992.

Nordenfelt, L., *On the Nature of Health*, Dordrecht: D. Reidel Publishing Company, 1987.

Nordenfelt, L., *Quality of Life, Health and Happiness*, Avebury, 1993.

Pickering, G., *Creative Malady*, London: Allen and Unwin, 1974.

Querido, A., "'The Wisdom of the Body': The Usefulness of System Thinking", in *Discipline of Medicine*, A. Querido, L. A. van Es, E. Mandema (eds.), North Holland, 1994.

Rosenfield, C., *From Beast-Machine to Man-Machine*, New York: Octagon Books, 1968.

Sheldon, A., "Toward a General Theory of Disease and Medical Care", in: *Systems and Medical Care*, A. Sheldon, F. Baker, C. P. McLaughlin (eds.), Cambridge, Mass: M.I.T. Press, 1970.

Wartofsky, M., "The Social Presuppositions of Medical Knowledge", in: J. L. Peset, D. Gracia (eds.), *The Ethics of Diagnosis*, Kluwer Academic Publishers, 1992.

Wulff, H. R., "Function and Value of Medical Knowledge in Modern Diseases", in: H. ten Have, G. K. Kimsma, S. F. Spicker (eds.), *The Growth of Medical Knowledge*, Kluwer Academic Publishers, 1990.

Wulff, H. R., "The Disease Concept and the Medical View of Man", in: *Discipline of Medicine*, A. Querido, L. A. van Es, E. Mandema (eds.), North Holland, 1994.

Wulff, H. R., Pedersen, S. A., Rosenberg, R. *Philosophy of Medicine*, Oxford: Blackwell Scientific, 1986.

KEVIN W. WILDES, S. J.

THE LOGICS OF MEDICINE: DISCOVERY AND EVALUATION

All too often in recent years, medicine and health care have been a locus of debate, confrontation, and discussion in contemporary secular societies. Medical ethics and health policy, along with questions about allocation of resources, have been areas of debate that have touched on both medical practice and hospital and public policy. Issues in bioethics have raised important moral questions that have been, and remain, controversial and contentious in secular societies. A great deal of energy, effort, and attention has been given over to these issues and questions. Indeed, much of the literature in bioethics has focused on specific questions such as abortion, euthanasia, physician-assisted suicide, or reproductive medicine. Each of these issues, as well as many others in bioethics, reflects diverse assumptions about the nature of medicine and appropriate professional medical practice. However, in the midst of all these discussions, there has been a remarkable failure to grasp the conceptual and philosophical problems underlying these moral controversies in secular societies.

In this essay, I would like to step away from these particular topics of bioethics that are often the focus of applied ethics and turn toward more foundational questions about medicine itself. This turn will help to explore what it is about medicine, as a practice, that elicits these types of medical questions and problems so important to bioethics. Part of the answer lies in the moral dimensions of medicine itself. Medicine is involved with crucial moments in the moral narratives of men and women. It touches on birth, sexuality, suffering, and death. This essay will argue that it is the nature of medicine itself that leads, in secular, pluralistic contexts, to the different views of appropriate or inappropriate medical practice. It will be argued that medicine involves two different logics. One is the logic of discovery and the other is the logic of evaluation. The two cannot really be separated, as the logic of evaluation sets the context for the logic of discovery. While one may draw a distinction between the two logics conceptually, that does not mean that they can be separated in practice.

To illustrate how these logics work in medicine, I will turn first to examples from the concepts of disease and the related process of diagnosis, and secondly to the use of outcome assessment and practice guidelines. With a more through understanding of the value-laden dimension of medicine, the

A.-T. Tymieniecka and Z. Zalewski (eds.), Analecta Husserliana LXIV, 49–56.

controversies of health policy and medical ethics become more understandable. Insofar as secular societies are places where different moral communities can meet, the medical enterprise will be best understood and interpreted in a variety of ways. In a morally pluralistic, secular society, even the work of the biological sciences may be controversial. The more morally diverse a society is, the less one can say about a professional ethics for medicine or a philosophy of medicine. In such contexts one will need to think of philosophy of medicine and medical ethics in the plural.

I. DISCOVERY AND EVALUATION

Medical knowledge and reasoning are complex, as the history of medicine bears out. In ancient Greek and Roman medicine, the disputes between the Rationalists and Empiricists were disputes about the justification of knowledge claims in medicine. Lester King points out that the medical theories often had an insubstantial quality about them. By the middle of the eighteenth century, a reaction began to set in, in which there was an interest in more defined areas of study and a greater regard for observation (King, p. 248). In efforts to make medicine more scientific physicians, like Thomas Sydenham (1624–1689), proposed a disciplined methodology of observation and treatment. However, Sydenham's proposal was criticized for its failure to account for observer bias. The late eighteenth and nineteenth centuries saw a fundamental reassessment of medicine and its claims of knowledge. There were important efforts to sort out the logic of medicine and to identify the fallacies that often plagued medical practice (e.g., excessive deference to authority, speculative reasoning without sufficient empirical observation). The work of F. R. Oesterlen (1812–1877) (*Medizinische Logik*) advanced criteria of inductive reasoning in medicine. Following Mill, Oesterlen viewed medical logic as the application of general logical principles to the field of medicine for the purpose of securing a coherent inductive and empirical science.

H. Wulff suggests that one can understand medical knowledge in a threefold division: biological medicine, clinical research, and clinical practice. Biological medicine's primary concern is with scientific research in biology. Clinical research is focused on the development of knowledge and technology in clinical medicine. Clinical practice involves realities of patients and diseases. A concern in philosophy of medicine has been to clarify each type of medical knowledge and the relationship of different areas to one another (Wulff, 1986). Certainly in the areas of clinical research and clinical practice, it becomes clear that medical knowledge is not based simply on a logic of discovery. The determination of what is best for the patient involves a dimension of evaluation as well as discovery. It involves values as well as facts. It involves the interpretation of the facts of the case.

Modern philosophy, since David Hume, has distinguished the realm of facts from the realm of values. In a famous passage from the *Treatise on Human Nature*, Hume wrote:

> But can there be any difficulty in proving, that vice and virtue are not matters of fact, whose existence we can infer by reason? Take any action allowed to be vicious: Willful murder, for instance. Examine it in all lights, and see if you can find that matter of fact, or real existence, which you call *vice*. In whichever way you take it, you find only certain passions, motives, volitions and thought (Hume, Bk. 3, Part 1, Section 1).

This position has not discouraged modern moral philosophy from attempting to discover morality elsewhere by introspection and intuitions. G. E. Moore, of course, brought the fact-value distinction to bear on ethics in his position on the naturalistic fallacy. Moore, of course, has argued that moral qualities are not part of the natural world of discovery. Philosophers of medicine, along with other philosophers of science, came to understand that facts are not simply data that are given. Facts, in isolation, are useless. They need to be understood in relationship to other facts. The relationship is what gives meaning and structure to the interpretation of the facts. Feyerabend, Kuhn, Lakatos, and Musgrave brought a deeper awareness of the sociology of knowledge, along with the role of cultural values and social constructs, to our understanding of the formation of facts. Philosophers of medicine such as Ludwik Fleck (1886–1961) and others of the Polish School of philosophy of medicine made important contributions to a philosophy of medicine, insofar as they explored the extent to which observations depend on the *apriori* ideas of the observers. Philosophers such as Fleck pointed out how facts are not simple brute realities to be interpreted, but rely on interpretation and evaluation to be discovered.

If one follows Wulff's threefold division of medical knowledge, the relationship of facts to values is justifiable. The role of values in the area of clinical practice is obvious to many. Indeed, the dimension of values, in relationship to facts, contributes to many of the issues in bioethics (e.g., treatment decisions). However, the relationship between facts and values (discovery and evaluation) is not present only in the realm of clinical knowledge. The logic of discovery in biological medicine is conditioned, in part, by prior commitments of evaluation. What do we seek to discover? Which facts are important for us to notice? While it is possible to distinguish the realms of discovery and evaluation, one recognizes the essential dependence of facts on values. This is not an ontological dependence, but an epistemological one.

II. DISEASE AND DIAGNOSIS

The logic of discovery is often thought in medicine to be essential to areas such as diagnosis. But diagnosis will depend on what is determined to be a disease in a society. Discovery, by itself, is not sufficient to give a complete account of medicine. Edmond Murphy has pointed out that what we call facts in the biological and medical sciences are bound up with our own preconceptions. A fact is, in the strict sense, an abstraction: that is, an incomplete selection of those things with which our senses come into contact; and it is not particularly surprising that others, making different selections, may attain different facts (Murphy, p. 60).

Murphy makes the helpful suggestion that we speak of artifacts rather than facts. He suggests the importance of the knower in relationship to the known. This suggestion is helpful when one turns to disease classification and diagnosis. There are a number of ways in which disease can be defined. However, a classification of disease (nosology) does not require precise definitions. Conceptions of disease have social implications beyond the domain of medicine. Diseases are often thought (1) to be non-voluntary; (2) to be organic; (3) to fall under physician expertise; (4) to fall below a socially defined minimal standard of accepted health (Veatch, 1973).

From the beginning, the fact of a disease will be related to certain value judgements. That is, the role of socially defined standards of health shape what counts as a disease and what does not. One cannot simply say that diseases are organic conditions, as not all organic conditions are treated as diseases. Male patterned baldness is an organic condition that may not be a disease. Until recently, infertility, as an organic condition, was not thought of as a disease. Diseases are identified because of social and cultural value assumptions and commitments. Standards of normalcy and deviance are defined in part by cultural contexts and goals. Disease, furthermore, is a particular subset of deviance. Some deviations are treated morally (e.g., criminal activity), while others are not. To classify a deviancy as a disease is to distinguish it from a moral failing. Disease is not seen to be directly under the control of the will of the individual. To classify a deviation as a disease is to direct social energies and resources towards it as well as to excuse the deviant.

Some have sought to identify physiological and psychological functions that are natural and essential to humans. Leon Kass and Christopher Boorse have argued that one can specify those functions which are integral to "being human." In this way we can develop accounts of disease. However, such views depend on assumptions about what is "natural" to humans. In contrast, others, like Joseph Margolis and H. T. Engelhardt, have argued that definitions of disease and health depend on culturally determined value-judgments, and as a result can only be understood in terms of particular cultures and ideologies. A culturally dependent account need not deny that there can be great similarities in what will count as disease across cultures.

Disease classifications then involve the conjunction of facts (certain empirical conditions) and values, insofar as they reflect social norms. The discovery of diagnosis depends upon a culture's view of disease. The logic of evaluation conditions the logic of discovery. In a recent issue of *The Journal of Medicine and Philosophy*, Roberto Mordacci argued that health and disease are part of a process of interpretation. That is, there may be common experiences, but how experiences are interpreted (as normal, healthy, or diseased) will depend on the cultural and social context.

In recent years, disease classification has taken on further roles with social and evaluational dimensions. Disease classifications have taken on new and important roles in rationalizing the bureaucracy of medicine and the use of medical resources. As Charles Rosenberg points out, disease classifications serve to rationalize, medicate, and legitimate relationships between individuals and institutions in a bureaucratic society. This function of disease classifications is nicely exemplified in third-party payment schemes, where the inchoate and possibly incommensurable experiences of individuals are transformed into the neatly ordered category of a diagnostic table – and thus made suitable for bureaucratic use. In this sense a nosological table is a kind of Rosetta Stone, providing a basis for translation between two very different yet structurally interdependent realms (Rosenberg, p. xxi).

III. OUTCOME STUDIES AND PRACTICE GUIDELINES

In recent debates about health policy and the allocation of resources, there has been an emphasis on "outcomes research" and practice guidelines as a way to respond to concerns about cost-containment and quality of care. "Outcomes research" involves statistical analyses of large patient populations to determine what methods work. Practice guidelines are then generated from the outcome data, and they serve as recommendations for medical practice.

However, as Fred Gifford has pointed out (Gifford, 1996), there are clear differences between the outcome data and the practice guidelines. The former are seen to be "factual" and "descriptive" (part of the logic of discovery) while the latter are considered to be more normative and evaluative. However, if one probes more deeply into the outcomes data and how it is assembled, one finds that a logic of evaluation shapes the process of discovery. There are assumptions in outcomes research about the levels of statistical significance, the non-objective concerns that are part of how variables are described, the value-laden dimension of what counts as a good or bad outcome. All in all, Gifford argues that ethical questions surround the epistemological ones.

In turning to the process of diagnosis, we find that it depends upon prior evaluational choices as to what counts as a disease. Even the study of medical outcomes depends, in part, on prior value judgments to determine what counts as a good or bad outcome. Discovery is conditioned by evaluation.

IV. THE PROFESSION AND PRACTICE OF MEDICINE

The exploration of diagnosis and disease classification makes it clear that the logic of discovery and the logic of evaluation cannot be separated in medicine. They condition each other. With this relationship in mind, one comes to more clearly understand why contemporary medical practice, in secular societies, is beset by controversies. In turn, one becomes skeptical of claims about a single, essential professional ethic in medicine. The relationships between facts and values, discovery and evaluation, lead to a re-evaluation of the profession of medicine within the context of a secular, morally pluralistic society.

Some have argued that the profession of medicine commits one to an ethic of medicine (Pellegrino and Thomasma) and that one can discover the nature of medicine. Indeed, medicine is often contrasted with the legal profession as a profession that is transcultural and transnational. To some extent this is true. But it is true for reasons other that those often given. Most people think that medicine cuts across communal boundaries because it is a science that simply deals with facts. However, it is clear that medicine is not simply applied biology. Rather, it is a complex social enterprise that involves prior value commitments about health and disease. While medicine works with facts – a logic of discovery – it also requires a logic of evaluation and interpretation. Facts need to be selected and interpreted. This is particularly true insofar as moral values are part of medical interpretations. How one understands health and disease will in turn influence how one conceptualizes medical practice. Henrik Wulff has argued that a biological or empirical model of illness often contributes to paternalistic medical practice. If concepts of health and disease can be fully understood in biological terms, then there may be no need to assign the patient an active role in the decision process (Wulff). If, however, determinations of health and disease are not just empirical concepts, but concepts that are also related to cultures and values, the patient and society will have a more active role in determining the burden of the disease and the extent of treatment.

In turning to particular issues in bioethics one can see how medicine can take very different approaches to facts. One can compare the different ways in which the medical profession treats people in the last month of their lives, to find extraordinary differences. One can also look at different cultural approaches to HIV. Some nations have an open, public confrontation with the disease and approaches to prevention. This disease involves views about homosexuality, drug use, condoms, education, privacy, discrimination, and sexuality in general. Medicine, like any practice, is embedded in a way of life (MacIntyre).

The more one understands the relationship of facts and values, discovery and evaluation, the more nuanced one will be in speaking about a philosophy of medicine, or a professional ethic for medicine. Those who argue for such an ethic often assume a biological model (as described by Wulff) or a model of common morality, in order to define the philosophy of medicine or the professional ethic of medicine. However, it becomes clear that the biological model cannot stand alone, as it is dependent on the evaluation of the patient's best interest. At the same time, the common morality model is difficult to sustain in secular morally pluralistic societies. In secular societies that endorse no particular view of the good life, medicine will more accurately be viewed as having many logics, philosophies, and ethics.

Kennedy Institute of Ethics, Georgetown University

REFERENCES

Boorse, Christopher. 1975. "On the Distinction between Disease and Illness." *Philosophical Review* **85**: 70–86.

Engelhardt, H. T. 1975. "The Concepts of Health and Disease." In H. T. Engelhardt and S. F. Spicker, eds. *Evaluation and Explanation in the Biomedical Sciences*. Dordrecht: Reidel, pp. 125–141.

Gifford, F. 1996. "Outcomes Research and Practice Guidelines: Upstream Issues for Downstream Users." *Hastings Center Report* **26**: 38–44.

Hume, D. 1978. *A Treatise of Human Nature*. L. A. Selby-Biggs, ed. Oxford: Oxford University Press, Clarendon Press.

Kass, Leon. 1975, "Regarding the End of Medicine and the Pursuit of Health." *Public Interest* **40**:11–42.

MacIntyre, A. 1983. *After Virtue*. Notre Dame, Indiana: University of Notre Dame Press.

Margolis, Joseph. 1976. "The Concept of Disease." *The Journal of Medicine and Philosophy* **1**: 238–255.

Mordacci, R. 1995. "Health as an Analogical Concept." *The Journal of Medicine and Philosophy* **20**: 475–497.

Murphy, E. A. 1979, "Classification and Its Alternatives." *Clinical Judgments: A Critical Appraisal*. H. T. Engelhart, S. F. Spicker, B. Towers, eds. Dordrecht: Reidel Publishing Co., pp. 59–85.

Pellegrino, E., Thomasma, D. 1981. *A Philosophical Basis of Medical Practice*. New York: Oxford University Press.

Rosenberg, C. 1992. "Framing Disease: Illness, Society, and History." In *Framing Disease: Studies in Cultural History*. C. Rosenberg & Janet Golden, eds. New Brunswick: Rutgers University Press.

Veatch, R. 1973, "The Medical Model: Its Nature and Problems." *Hastings Center Studies* **1**: 64.

Wulff, Henrik. 1986. *Philosophy of Medicine*. Oxford: Blackwell Scientific Publications.

JAN DOROSZEWSKI

AN ANALYSIS OF MEDICAL KNOWLEDGE AND REASONING

INTRODUCTION

Until relatively recently, the human organism was always considered as an entity that was extremely complex, difficult to comprehend (if intelligible at all) and not subject to deterministic laws. At the turn of the nineteenth and twentieth centuries, thanks to the progress of science, a hope appeared that things were not so bad, that we could understand the human organism, which started to be seen as a precise, if not quite regularly working, mechanism. In the second half of the twentieth century, however, it has become evident that the living organism cannot be viewed in this way, and thus certain aspects of the previous, traditional way of looking at the organism have started to attract interest once again. This time, nonetheless, its complexity and indeterminacy have appeared in a new light, and it is the very same scientific progress that has now started to suggest how the components of the organism (and the way they work) may be perceived. In order to deal with the complexity and uncertainty, new sciences have appeared, such as systems theory, cybernetics and others (including the youngest of them, the theory of chaos). Research in the field of goal-directed human behaviour and efficient action (decision theory, praxiology and theories of management) is also beginning to bring results that are relevant to medical theory and practice. As regards medicine, perhaps even more crucial is the fact that the same disciplines in which a few score years ago reductionist, over-simplified tendencies dominated (e.g., physiology, biochemistry, pharmacology) are now becoming more mature – and seem to be ready to produce ideas that are sufficiently deep and broad to adequately describe extremely complex phenomena. A prerequisite, however, for the fulfilment of such a possibility is the development of a general methodological theory of these sciences.

The present paper is divided into three parts. In the first part, remarks on the general structure of biomedical knowledge (rather than medical knowledge as a whole) and its main types are presented in a manner resembling the cognitive science approach. In the second part, a general description of medical action is present: it is to some extent influenced by a praxiological viewpoint. The third part contains a presentation of a study of biomedical knowledge using a logical apparatus.

A.-T. Tymieniecka and Z. Zalewski (eds.), Analecta Husserliana LXIV, 57–66.

The papers and books that most influenced my research, part of which is presented in this paper, are those of Tytus Chałubiński[1] and Władysław Biegański[2] as regards medicine, and – from other disciplines – Tadeusz Kotarbiński[3] and Kazimierz Ajdukiewicz.[4]

The last introductory remark is the following: The structure of knowledge, which – according to the title of this paper – is the principal theme, may be studied from various points of view and for different purposes. I am interested mainly in the dynamic aspects of the structure of knowledge, i.e. those that are directly linked with the use of knowledge in reasoning and for purposes attainable by reasoning. Therefore, I will concentrate on the structure of medical knowledge as applied in the process of problem-solving.

I. MEDICAL KNOWLEDGE AND PROBLEMS

Medical action is a kind of goal-directed behaviour in which reasoning is inseparably linked with physical activity. In other words, it is based on formulating and solving problems which have a mixed mental and physical character. As the term "problem" is used in different senses, it may be appropriate to make clear that by this notion I understand here an intellectual task which requires non-trivial reasoning for its fulfilment.

Every problem arises against a certain psychic and intellectual background. Medical problems, with which a physician is confronted, are always understood by him (her) in connection with his (her) concept of the physician's fundamental task, duties and vocation, i.e. on a general and varied knowledge rooted in his education and his ability to observe facts.

Problem processing[5] constitutes a mental background for all of a physician's actions, and is their leading force. The solution to a medical problem consists in the formulation of a plan (design) of diagnostic and therapeutic action in a given case and/or the evaluation of the supposed result of an action (diagnostic or therapeutic). The problem-solving process begins with an initial set of statements (*initium*) describing what is already known and stating what is to be found out (*explorandum*), and leads to the formulation of the solution (*exploratum*).

Using logical terminology, problem-solving reasoning is based, on the one hand, on statements describing what the subject already knows about a given situation and, on the other hand, on more abstract statements that are part of a general knowledge of a certain kind. Mental operations, being the basis of the problem-solving process, consist of generating statements (sentences, expressions, including directives) constituting the *exploratum*, by deriving them (deducing in a broad sense) from the general knowledge and particular statements related to the initial concrete knowledge.

As regards medical problems, the initial particular statements describe two types of factors. Firstly, the condition of health of the patient and his or her other features as they appear to the physician during the first encounter, e.g. his or her psychological background, individual values and desires, goals and designs, etc. Secondly, the various elements of the surrounding situation, such as accessible diagnostic and therapeutic means, transport facilities, the possibility of communicating with other people (e.g., other doctors), etc. These factors are usually quite evident and therefore do not need to be expressed in the form of complete propositions (which, however, are necessary for a theoretical analysis). Statements of the type: "This patient complains of such and such an ailment" or "The present situation is so and so", are true only in a given unique case; in contradistinction, general statements are true in every case that belongs to a certain class. In the process of problem-solving, particular propositions are linked with general ones, i.e., those of which a general knowledge is composed.

In general empirical knowledge, three main types may be distinguished: object-directed (sometimes called declarative), operational, and instrumental ones. Abstract or formal knowledge (such as mathematics or axiology) cannot, of course, be divided in a similar manner.

Object-directed knowledge describes sets (classes) of objects from various points of view and the features of these objects that are of interest to a given discipline. The object-directed knowledge that is utilised in medical problem-solving is: anatomy, physiology, biochemistry, pathology, psychology, etc. A typical expression of a statement belonging to object-directed knowledge is a declarative sentence.

Operational knowledge is related to the methods leading to the attainment of various goals (cognitive and practical); operational statements describe then the methods of action, especially the sequences of procedures, including bodily movements, performed for the attainment of a certain goal (e.g. diagnostic or therapeutic). Typically, operational statements are formulated in the imperative mood (instructions, advice, directives).

By the term "instrumental knowledge" I understand sets (systems) of statements that constitute linking elements between object-directed and operational types of knowledge. They reflect the laws on which the methods used in a certain field, e.g., diagnostic or therapeutic, are based; in other words, they present the rationale of methods. The linguistic form of instrumental statements is similar to the linguistic form of those belonging to object-directed knowledge, i.e., the declarative mood.

I am going to propose the following model of problem-processing from the viewpoint of how the various kinds and types of knowledge distinguished above participate in it. In order to present the problem in the right way, we need to distinguish two things. The first is the set of both particular and general statements which doctors use as premises in their reasoning. The second is all other components of problem-solving reasoning. I will call it a "problem-directed piece of knowledge" or, for short, "problem knowledge". It is composed of particular and general statements. The former describe features of the already known condition (in the broad sense) of the patient and of the situation; the latter are elements of general knowledge (object, operational and instrumental) which are extracted from it for a given purpose.

Problem-solving reasoning consists in operations performed in the frame or on the basis of such a piece of knowledge (with some modifications, as necessary). The expected solution (*explorandum*) to the problem is a set of statements (accepted if there is a satisfactory degree of certainty) describing the condition of the patient, and a set of directives related to the treatment of the patient.

An example may be helpful to illustrate the above idea. For the sake of clarity, I have chosen a very simple medical problem and described the whole process of inference-making in a slightly simplified manner.

Example I

A patient who complains of a strong pain in the forearm after an accident comes to an out-patient department; the radiogram indicates signs of a fractured radius. The doctor's problem, then, is this: *The patient has a broken radius; what treatment is indicated*?

The *general motivational background* of the doctor is: his or her professional desire to heal the patients (to fulfil their health-related wishes).

The *concrete knowledge* of the doctor is:

a) Concerning the patient's condition and attitude: This patient's radius is broken, this patient wants to have his forearm healed.
b) Concerning the situation: The doctor is able to apply a plaster dressing (i.e., plaster of Paris is available at the surgery and the doctor knows how to make such a dressing).

In this situation, the *general object-directed knowledge* may be described by the following statements:

– If a bone is broken but is otherwise normal, then the bone healing process makes healing of the fracture possible.

- If a bone is broken and the healing process is going on and the broken fractures are in uninterrupted contact, then the integrity of the bone will be restituted.

The *instrumental knowledge* may be summarised as follows:

- If a bone is broken and immobilised and set, then the broken fractures are in uninterrupted contact; if a bone is broken and is immobilised and set, the fracture will be healed.

Some statements typical for a general *operational knowledge* are the following:

- If a plaster dressing is applied to the part of the body where the broken bone lies, the broken bone will be immobilised.
- If anybody wants to apply a plaster dressing to a patient's leg (forearm, etc.), he or she should perform the following operations; if and only if the following operations (...) are performed, the leg (forearm, etc.) shall be immobilised and set.
- If a plaster dressing is applied to the broken bone of a leg, the bone's integrity will be restituted (the fracture will be healed).

From the above propositions, concrete *instrumental statements* are derived:

- If a plaster dressing is applied to this patient's forearm, the fracture of his radius will be healed.
- If the patient wants to have his forearm healed, the plaster dressing should be applied to his forearm.

The *solution to the problem*, then, is the following: A plaster dressing should be applied to the forearm of this patient.

Once the decision is made, a piece of operational knowledge describing how to make a plaster dressing is applied to the treatment of this kind, for a particular patient and his or her forearm.

II. A MEDICAL PROBLEM-SOLVING ALGORITHM

The above description mainly concerns the basic parts of a problem-oriented piece of knowledge, while the problem-solving process is only mentioned in a very superficial way. A detailed analysis of this process would be impossible in a short time; instead, then, a general outlook may be presented in an algorithm-like form (Fig. 1).

The starting point of the process is what is already known about the patient and the situation. The physician considers whether his or her knowledge is sufficient to decide what treatment should be performed in a given case. If yes, he or she proceeds to therapeutic action; if not, diagnostic action begins.

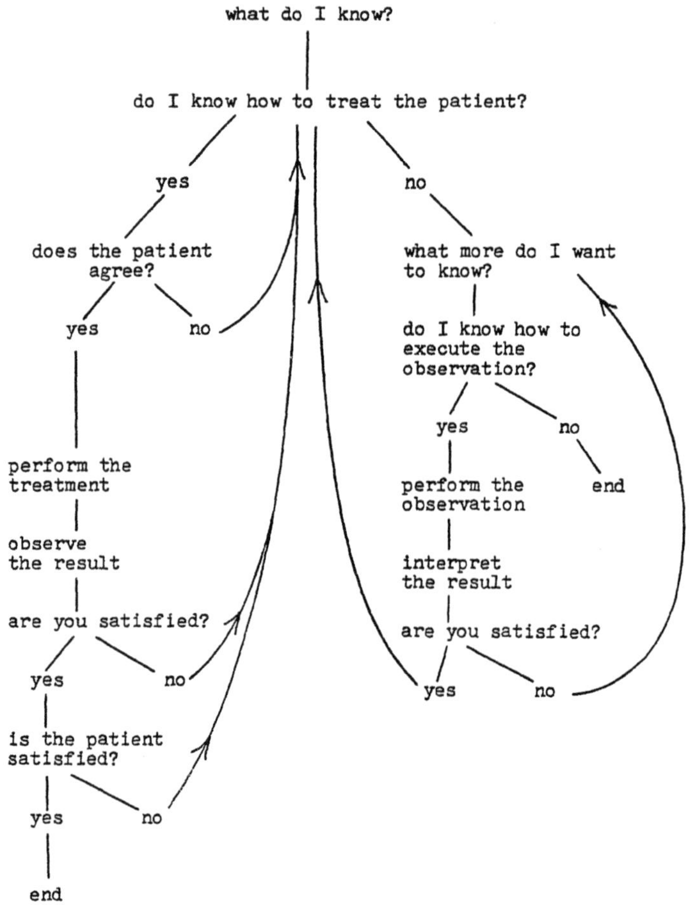

Fig. 1

The therapeutic sequence of actions starts with the presentation of the doctor's plan to the patient. If agreement is reached, treatment is performed, and then its results are observed, and if both doctor and patient are satisfied, the action is terminated. If the patient does not agree to the doctor's proposal or if the results of the treatment are not satisfactory, the question of whether or how to continue the patient's treatment arises once again.

If the diagnostic line of action is chosen, the physician first decides what more he or she wants to know and whether he or she is able to carry out the necessary examination. If yes, the examination is performed and its results are interpreted. If the desired information has been obtained, the doctor comes back to the question of how to treat the patient. If the results of the examination are insufficient, the question of what more needs to be known arises again.

Of course, this scheme is an evident over-simplification resembling "the shadow of smoke" as the Greeks would say. It shows some chosen, artificially isolated factors which, in reality, form large sets of interrelated elements; it tries to reflect a multidimensional structure of notions and propositions on a flat surface; extremely complex and diversified mental and physical operations are uniformly indicated by arrows. Which course of action the doctor takes is a matter of choice. However, one feature of the process of medical problem-solving is faithfully represented in this scheme, namely the fact that mental operations are inseparably intermingled with physical ones, the former leading to the latter and the latter initiating the former.

III. A LOGICAL ANALYSIS OF KNOWLEDGE

In order to examine in more detail the logical structure of medical knowledge in a certain discipline or in a problem-oriented trans-disciplinary piece of knowledge, an approach based on the analysis of the form and contents of the notions (terms) and statements (propositions, sentences) may be applied. Since natural professional language is, on the one hand, the main "carrier" of medical knowledge, and on the other hand, has a certain invariant structure and is closely linked with reasoning processes, an analysis of original linguistic expressions, especially written texts, seems to be appropriate for such a study. This approach may be applied to various types of knowledge, such as the above-mentioned object-directed, operational and instrumental types of knowledge and different disciplines (morphological, physiological, pathological, epidemiological, etc.) as well as to the processes of inference-making. Although it is based on general logical principles, in its present form it is best suited to the study of sciences, including medical science. The application of this approach to other domains, e.g., to psychology or physics, would probably necessitate certain modifications and developments. On the other hand, I cannot see clearly whether and how it could be adapted for the purpose of the study of non-empirical sciences, e.g., ethics.

In order to present the basic features of this approach, I will come back to a statement cited in the first part of this paper as an example of object-directed knowledge used in problem processing. In order to illustrate different aspects of its contents as concisely as possible, the statement I am going to examine is not an original textbook one; however, it has all the features of a genuine expression.

Example II

The statement is as follows: if a bone is broken but is otherwise normal, then the bone healing process is going on, making healing of the fracture possible.

The first step of the analysis is a contents-equivalent (as I call it) reformulation of the statement which consists in making it formally simpler (even though this may make it somewhat clumsy from the stylistic point of view). This is accompanied, if necessary, by clarification and/or a certain enrichment of its contents. This is, by the way, a unique step which requires a certain amount of professional (in this case object-directed) knowledge from the person doing the analysis.

The above statement may be reformulated in this way for every bone that is part of a living organism: when a normal bone is broken, separated bone fragments appear and new bone tissue is formed nearby; after some time, its formation usually causes the bone fragments to link together.

What is the meaning of the above sentence? Its particular parts describe:

1) objects (organism, bone, bone fragments, new bone tissue);
2) properties (healthy, broken, appearance, formation);
3) spatial relations (separation, linking together, nearby, being part of);
4) temporal relations (when, after some time);
5) modality (usually);
6) second order relations (causes);
7) quantification (every).

In order to show how the corresponding expressions appear in the sentence, it may be rewritten, using various ways of distinguishing the words and phrases; those that mean objects are indicated by a bold font, properties by italics, spatial relations by bold italics, temporal relations by capital italics, modality by bold-font capitals, second order relations by capitals, quantification by capital letters. The sentence, then, looks as follows.

For EVERY **bone** in an **organism**: *WHEN* a *normal* **bone** is *broken*, ***separated*** **bone fragments** *appear* and **new bone tissue** is formed ***nearby***; *AFTER SOME TIME* its *formation* **USUALLY** CAUSES the ***linking together*** of **bone fragments**.

Taking into condideration the meaning of some of the above-mentioned parts of the statement, a graphic representation may be used; it may show the

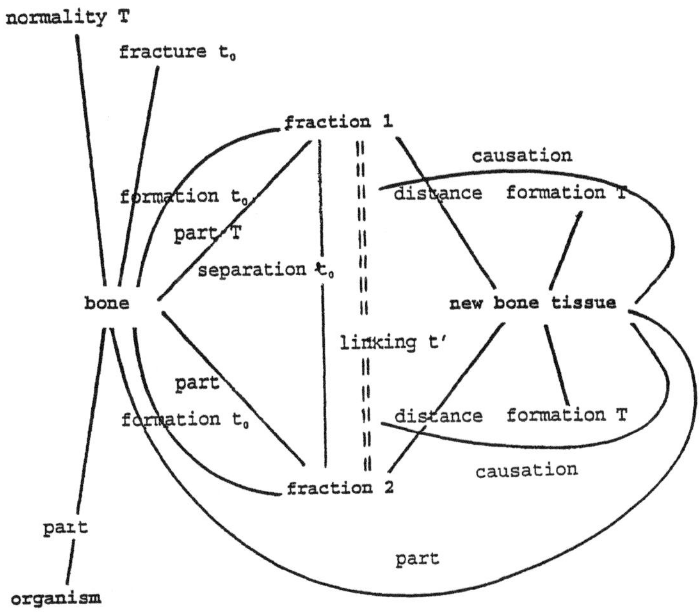

Fig. 2

result of the analysis somewhat more clearly (Fig. 2).

In one-argument relations, (Fig. 2) properties (as normality, fracture, formation) are represented by straight lines without description. Two-argument relations are indicated by word comments: the bone is part of the organism, the bone fragments are parts of the bone, new bone tissue is near the bone fragments, the bone fragments are separated, etc. A sinuous line means "usually linked", i.e., a modal relation, the simplest interpretation of which is: " is linked or is not linked". A second-order relation of causation is depicted by lines connecting the property of "forming new bone tissue" with the relation of "usually causes linking together of the bone fragments". The time factors are indicated to convey the meaning of events (t, t') and a period (T) between the events.

In a similar manner, various types of statements may be analysed, including instrumental and operational ones. As a result, a network of interconnected names of objects, properties, relations, etc. may appear; e.g., a whole problem-oriented piece of knowledge may be thus represented. Sometimes such an analysis leads to an extremely complex result. Its complexity, however, is balanced by clarity and the possibility of performing various inference-making operations in a verifiable manner. A graphic presentation makes this easier to understand.

Effective problem-solving requires, in almost every case, that inference-making be conducted on the various levels of generality of problem-oriented knowledge. The notion of the level of generality is based on the relationships between the sets of designates, the names of objects and predicates (denotation). The broader the denotation of terms which appear in a given piece of knowledge, the higher its level of generality. Coming back again to the above-analysed example of the problem concerning the treatment of a bone fracture, the denotation of notions in ascending order is the following: fracture of radius in a given patient, radius fracture, bone fracture, wound (understood as disruption of tissue continuity), trauma; other notions involved in this piece of knowledge (e.g., for healing these conditions) may also be similarly arranged. Thus, a piece of knowledge describing the treatment of a radius fracture is lower in the hierarchy of generality than a piece of knowledge which describes the treatment of a fracture of every bone, which is less general than that of healing a wound, that of the treatment of trauma, etc. Inference-making performed only on a certain level of generality is much less effective than multi-level reasoning, and it is the latter method which is used in natural thinking. It is of paramount importance, therefore, that such types of problem-solving processes be studied in detail, using – among other methods – a logical model. The main obstacle to understanding these problems is connected with the fact that the pieces of knowledge on various levels of generality do not reproduce on a different scale in an identical structure. In other words, the lower level knowledge cannot be simply deduced from the higher level one. It seems that the above-presented method for studying the logical structure of medical knowledge may also prove useful for this purpose.

Medical Centre of Postgraduate Education, Warsaw

NOTES

1 T. Chałubiński, *Metoda wynajdywania wskaza/lekarskich. Plan leczenia i jego wykonanie* (A method for arriving at medical indications. Plan of treatment and its execution), Warszawa: Gebethner and Wolf, 1874.

2 W. Biegański, *Logika medycyny, czyli krytyka poznania lekarskiego* (Logic of medicine, or critical analysis of medical cognition), Warszawa: Wende, 1908.

3 T. Kotarbiński, *Traktat o dobrej robocie* (A treatise on good work), Wrocław: Ossolineum, 1982.

4 K. Ajdukiewicz, *Logika Pragmatyczna* (Pragmatic logic), Warszawa: PWN, 1965.

5 See: J. Doroszewski, "Structure and Solving of Medical Problems: A Unifying Approach", Proceedings of the First International Congress of Medical Philosophy, 30 May to 4 June 1995,

JOACHIM WIDDER

THE SIGNIFICANCE OF "ALTERNATIVE MEDICINE" FOR THE PHILOSOPHY OF MEDICINE*

In this paper I will first briefly discuss the concept of effectiveness, then turn to the motives of patients for using alternative medicine. Next, the three main classes of factors usually isolated in the healing process will be considered, and finally, some examples will be given for illustration.

GOALS OF MEDICINE AND EFFECTIVENESS OF METHODS

Effectiveness is increasingly being used as a core concept for measuring different degrees of the quality of medicine, including the provision of health care, as well as for the assessment of different treatment options.[1] The term "effectiveness" is here used as meaning the power to produce a certain event; in other words, the power of a medical (or health-care) procedure to achieve certain goals.

Although there is increasing talk about effectiveness in medicine and health care, there seems to be no sufficient analysis of the concept of effectiveness itself, and no sufficient debate about the goals of medicine, without which the term effectiveness must remain vague. The goals are still frequently considered as more or less self-evident and out of dispute by medical practitioners, in spite of major disagreement in the bioethics literature in the last three decades and in spite of a growing supply of and demand for alternatives to orthodox (Western) medicine. Furthermore, effectiveness is not always discussed together with the question of how important an even effectively achievable aim is in relation to other aims, and which means are justified by a certain end.[2] For instance, can helping a patient mobilize his or her "inner energy against disease" sometimes be a higher aim than the attempt to prolong life? Are there situations where intentional killing can be an appropriate way to relieve suffering? Should measures sometimes be taken to force persons exhibiting a health-damaging behaviour to stop that behaviour? Could helping to live through or with an illness sometimes be an even more important goal of medicine than immediate and complete symptom relief? And so on.

It is clear that effectiveness is a concept describing the degree to which a method realizes a defined goal. On the other hand, alternative systems of medicine are often criticized because of the presumed ineffectiveness of their

A.-T. Tymieniecka and Z. Zalewski (eds.), Analecta Husserliana LXIV, 67–74.

treatments. However, it is absolutely meaningless to compare the effectiveness of different methods if they do not relate to the same goal. This, though trivial, nevertheless needs to be mentioned in the debate about orthodox versus alternative (or complementary) medicine, because it seems obvious that oftentimes when a certain kind of alternative medicine is called ineffective, this actually means that it pursues strange goals or at least that there is something wrong with the hierarchical arrangement of its goals of treatment.

MOTIVATIONS FOR CONSULTING ALTERNATIVE MEDICINE

Frequently, alternative medical systems are viewed as more or less well defined sets of medical procedures: pricking certain points in traditional Chinese medicine; using medicines that are produced in a certain way in homeopathy, etc. The medical tools used are sometimes thought to be the most distinctive feature of a kind of alternative medicine. However, when one takes seriously the motives of people (patients and physicians) in using various kinds of non-orthodox medicine, a quite different picture emerges:[3] They are interested in a more holistic understanding of health and disease, they are more actively interested in trying everything possible to advance health, and they believe that orthodox treatments entail a greater risk of harm while alternative medicine would mobilise the body's natural ability to heal itself. Furthermore, they attribute a high priority to the quality of the relationship between patient and physician. The doctor's time and attention are valued very highly and, most notably, the expectation of a cure is secondary to the interest shown by the practitioners in the patients' views of their problem. Most *users* of non-orthodox medicine use it as a complement to orthodox medicine rather than as a substitute.[4]

It is obvious that quite a few of these motives given by users of non-orthodox medicine deal with the conceptions of health and disease that people attribute to different kinds of medicine, and with the perceived quality of the medical encounter. Physicians, on the other hand, notoriously tend to underestimate their patients' desire to understand their condition, and to literally come to terms with disease and with how they should think about healing in principle. It seems to be untrue that most patients always just want to get rid of symptoms and disease.

THE HEALING PROCESS

By way of briefly sketching the three components which are known to influence the healing process, I want to illustrate the thesis that the effectiveness of alternative medicine compared with the effectiveness of orthodox Western medicine cannot be discussed in a meaningful manner without referring to the different conceptions of disease and healing embedded in different anthropologies of the medical systems in question. Any system of medicine rests on some more or less explicit conception of normality whereupon health and disease are understood, and from which in turn the basic principles of how healing could come about are derived. For instance, if disease is thought to consist in abnormal cell or organ function, cells and organs will be the primary target of healing interventions; if disease is thought to consist in an imbalance of the spiritual vital forces – as is the case in homeopathy – these spiritual forces will be the primary target of healing interventions.

Traditionally, three main components are isolated in the healing process from Chałubiński, Biernacki, and Biegański to contemporary writers:[5]

(1) self-healing,
(2) the influence of non-specific therapeutic factors,
(3) the influence of specific therapeutic factors.

(1) Josef Dietl (1804–1878), a nineteenth-century physician who began his career in the Vienna Medical School and then became professor in Kraków, as one of the best known advocates of the so-called "therapeutic nihilism,"[6] became famous for his opinion that "only nature can heal." Physicians should therefore be honest enough, he says, to confess that refraining from therapeutic activity while predicting the course of the disease would be the best help they could offer their patients. Physicians should console and reassure or comfort patients with respect to their prognosis. Otherwise, they would do more harm than good.[7] This seems to be a clear example of a system where self-healing is seen as the only factor in the healing process to be taken seriously.

(2) Strictly speaking, however, what Dietl means can only be called self-healing, if health and disease are regarded as pure biological or natural events, because as soon as a patient physician-relationship is established, then various non-specific factors are brought up and the question about the purpose of medicine inevitably arises. Dietl is aware of this, but regards psychological factors as not belonging to medicine proper. Mostly these non-specific factors are conceived of as psychic factors. The quality of the personal relationship as well as the therapeutic setting become involved here, and part of the so-called placebo effect must obviously be attributed to the influence of this emotional dimension (I will give an example later). But there is also a more cognitive side to this category of non-specific therapeutic factors which consists in the perception of disease, health, and the healing

process itself, by the patient as well as by the physician. The conceptualization of these basic medical categories and their meaning for humanness will influence both the physician's and the patient's expectations and behaviour, and the disease and symptoms directly – if, and only if, disease is not defined as a purely biological event. We already saw the importance of this issue in studies of the motivation of patients to consult non-orthodox medicine, where not only psychic factors concerning the patient-physician relationship but also cognitive aspects concerning the conception of disease were involved. To give a short example: In anthroposophic medicine which incorporates parts of homeopathy, the reason for being very reluctant to immunize children against childhood diseases is seen in the importance of those diseases for the development of the personality of the child. Obviously, for orthodox medicine such a conception of diseases is alien, as it can tell the risk of childhood diseases, but does not attribute a value for the development of the personality to these diseases. This is a quite extreme example, but it shows nicely that the conception of disease both by patient (or parent) and physician may heavily influence what will happen in the medical encounter and which aims will be pursued.

All these categories of non-specific therapeutic factors – the psychic dimension, the therapeutic setting, and the cognitive conception or theory of disease – always include heavily subjective components. This means that they irreducibly depend on the very persons acting in and affected by the therapeutic process.

(3) This difficulty does not arise when specific therapeutic factors are considered. These are defined as means to a certain aim, and the term specificity implies the describability in objective terms of the aim. Undoubtedly one of the most prominent methodological presuppositions of modern Western medicine consists in the endeavour towards specification and therewith making the whole therapeutic process objective. The classical randomized case-control clinical trial is the well established method to study specific factors and it is designed to systematically neutralize the potential influence of all non-specific factors in the healing process. It has to be admitted, however, that the aims of therapy, objectively describable as they may be, are at last never justifiable independently of subjective valuation, *if* medicine is to leave at least some room for personal freedom to decide upon one's own way of dealing with illness. This qualification seems to be necessary if person-independent justice is not to be the most fundamental category of medicine, but health and healing. But if this is conceded, then the argument for regarding the non-specific factors as non-essential in the healing

process – because of their subjectivity – becomes very weak. Rather, it seems necessary to study in much greater detail the workings of non-specific factors in the healing process. The growing tendency of patients to consult and of physicians to offer several kinds of alternative medicine seems to be a clear indication of orthodox medicine's failure to take seriously precisely those factors in the healing process that it normally takes pains to exclude when effective means of therapy are to be established.

EXAMPLES

There are still very few studies directly addressing the question of how non-specific factors influence the therapeutic process and how they interact with specific measures. I will briefly consider some of them.

A study in the United States investigating the effect of drug administration on analgesic response[8] shows that if a saline solution without a specific drug is infused, this infusion will alleviate pain only if analgesia is expected by the patient – this is the well known placebo effect. This placebo effect is antagonizable by the morphine-antagonist naloxone. But if one looks at this study from a different angle[9] and asks the question: what is the effect of a naloxone infusion in patients experiencing pain?, one gets an interesting result: if naloxone is given to patients who do not expect to receive pain medication, it slightly (although statistically not significantly) reduces pain, but if it is given to patients expecting to receive pain medication, it increases pain. Thus, the effect of naloxone is turned from positive to negative if the expectation of the patient changes. This experiment gives an example that an effect of a drug is not only aggravated or lessened, but can sometimes even be inversed by non-specific factors.

Another study in France investigated the effect of informed consent on the analgesic activity of placebo and a pain killer in cancer pain.[10] A pain killer (naproxen 500 mg) as well as a placebo were more effective when information (about the verum) was given and consent obtained in either group of patients: i.e., those who expected pain relief were more effectively relieved from pain than those who did not expect pain relief. Moreover, placebo with consent was a more effective pain killer than the drug given without informed consent. This result again indicates a considerable interaction between non-specific and specific therapeutic factors. It is a very clear example that the effect of a drug may heavily depend on several psychic and cognitive factors and not only on biological mechanisms. In various other studies investigating the influence of informed consent on the overall effect of a therapy, it is always observed that information and consent do change the outcome of therapies.[11] Valid consent as a requirement based on ethical reasons gets to interact with procedures based on other reasons, for instance biological or pharmacological reasons. It is evident in this example that the decision about which kinds of ingredients in the therapeutic process one favours is a decision between different images of humanness. A "positive effect" from the biological perspective can be a "negative effect" from the ethical perspective and *vice versa*. How these perspectives should be integrated is a question of normative anthropology,[12] and this interaction of specific with non-specific factors raises questions concerning the daily, routine use of treatments tested in sophisticated clinical trials.[13] What one considers as medical knowledge proper (as opposed to peripheral or non-essential knowledge) will depend on which kind of effects one regards as most important.

This also applies for the next example: In a study in Denmark investigating the effects of acupuncture in moderate, stable angina pectoris,[14] one can see that much depends on the way in which specificity and non-specificity are defined within a therapy. When patients with angina received acupuncture treatments, they experienced an increased exercise tolerance and a delay in time to onset of pain during exercise (ECG parameters were not significantly altered), they needed significantly less nitroglycerin, and the angina attack rate decreased (also significantly). The control-group of patients received sham acupuncture: i.e. needles were inserted in the same dermatomes as in real acupuncture, but outside the Chinese meridian system and not at defined acupoints; needle sensation as required in traditional Chinese medicine was not elicited. The result was that sham acupuncture exhibited the same positive effects on exercise tolerance, nitroglycerin consumption, and angina attack rate. It must be noted that all other elements of treatment with the exception of pricking the "real" points and eliciting the needle sensation were the same in both study groups – and both groups profited from the treatment. The physician-patient relationship, the manner of talking about disease, the selecting of points on the body surface, the insertion of the needles and touching the patient's body, and the whole atmosphere, was shaped by traditional Chinese medicine. This study seems to suggest that acupuncture as well as sham-acupuncture treatment may be effective to increase a patient's well-being and exercise tolerance in moderate angina pectoris. A positive effect on exercise tolerance has not been shown by drug placebos,[15] suggesting some different dimension than a classic placebo effect. This different dimension would have to be investigated by studies comparing acupuncture and sham-acupuncture versus other kinds of treatment, of course. The study also shows, however, that the decisive element in acupuncture in moderate angina does not consist in pricking the "right" acupoint, nor in the needle sensation, but in something else. If one concludes from a study like this that acupuncture is ineffective, one implies that its demonstrated positive effects on well-being do not count as medical treatment. This study therefore does not show that acupuncture is ineffective, but suggests that acupuncture should reconsider its consistency at least for the indication of moderate angina pectoris.

CONCLUSIONS

These examples could only hint at the potential importance of studying in much greater detail non-specific factors of therapies and the interaction of specific with non-specific factors. Even heavily subjective experiences and valuations could be studied, but new and more refined study-designs would have to be developed to achieve this. The randomized (blinded) clinical trial is an ideal instrument for studying the specificity of a therapeutic factor, but it is inappropriate for the study of non-specific factors in therapies, because it is based on their intentional neutralization.

Moreover, the problem of the goals of medicine should not be studied only from a purely ethical perspective, where the so-called "facts" are provided by (orthodox Western) medicine with the preferences then judged by ethical reasoning, because in such a framework the "facts" are already imbued with the implicit anthropological assumptions of modern scientific medicine with its tendency towards explanation of immediate experiences by biological processes. A philosophical analysis of the anthropological underpinnings of different systems of medicine should at least accompany ethical considerations in medicine.[16] This would be especially helpful for the debate about alternative medicine.

Vienna University Hospital

NOTES

*Supported by the Austrian Research Fund (Proj Nr P 09839-MED).

[1] Mooney, G. "The Demand for Effectiveness, Efficiency and Equity of Health Care." *Theoretical Medicine* 10, 1989, pp. 195–205. Relman, A. S. "Assessment and Accountability: The Third Revolution in Medical Care." *N Engl J Med* 319, 1988: 1220–2. Ellwood, P. M. "Shattuck Lecture – Outcomes Management: A Technology of Patient Experience." *N Engl J Med* 318, 1988: 1549–56. Roper, W. L., Winkenwerder, W., Hackbarth, G. M., Krakauer, H. "Effectiveness in Health Care. An Initiative to Improve Medical Practice." *N Engl J Med* 319,

1988: 1197–1202. Berwick, D. M. "Continuous Improvement as an Ideal in Health Care." *N Engl J Med* 320, 1989: 53–6.

[2] Veatch, R. M. "The Role of Ethics in Quality and Accountability Initiatives." *Medical Care* (Supplement) 33, 1995: JS69–JS76. Gifford, F. "Outcomes Research and Practice Guidelines. Upstream Issues for Downstream Users." *Hastings Center Rep* 26 (No. 2), 1996: 38–44.

[3] Schär, A., Messerli-Rohrbach, V., Schubarth, P. "Schulmedizin oder Komplementärmedizin: Nach welchen Kriterien entscheiden sich die Patientinnen und Patienten?" *Schweiz Med Wochenschrift* 124 (Suppl. 62), 1994: 18–27. Murray, J., Shepherd, S. "Alternative or Additional Medicine? An Exploratory Study in General Practice." *Soc Sci Med* 37, 1993: 983–8. Moore, J., Phipps, K., Marcer, D. "Why Do People Seek Treatment by Alternative Medicine?" *Brit Med J* 290, 1985: 28–9.

[4] Eisenberg, D. M., Kessler, R. C., Foster, C., Norlock, F. E., Calkins, D. R., Delbanco, T. L. "Unconventional Medicine in the United States. Prevalence, Costs, and Patterns of Use." *N Engl J Med* 328; 1993: 246–52. Ernst, E. "Complementary Medicine. Scrutinising the Alternatives." *Lancet* 341, 1993: 1626.

[5] Biegański, W. *Medizinische Logik. Kritik der ärztlichen Erkenntnis.* Würzburg: Kabitzsch, 1909, 218–237. Klijnen, J., de Craen, A. J. M., van Everdingen, J., Krol, L. "Placebo Effect in Double-Blind Clinical Trials: A Review of Interactions with Medications." *Lancet* 344, 1994: 1347–9.

[6] Wiesemann, C. *Josef Dietl und der therapeutische Nihilismus. Zum historischen und politischen Hintergrund einer medizinischen These.* Frankfurt/M: Lang, 1991.

[7] Dietl, J. "Praktische Wahrnehmungen nach den Ergebnissen im Wiedner-Bezirkskrankenhause." *Zeitschrift der k.k. Gesellschaft der Aerzte zu Wien* 1845, 1 (2 Bd): 9–26.

[8] Levine, J. D., Gordon, N. C. "Influence of the Method of Drug Administration on Analgesic Response." *Nature* 312, 1984: 755–6.

[9] Klijnen, J., de Craen, A. J. M., van Everdingen, J., Krol, L. "Placebo Effect in Double-Blind Clinical Trials: A Review of Interactions with Medications." *Lancet* 344, 1994: 1347–9.

[10] Bergmann, J. F., Chassany, O., Gandiol, J. et al. "A Randomised Clinical Trial of the Effect of Informed Consent on the Analgesic Activity of Placebo and Naproxen in Cancer Pain." *Clinical Trials and Meta-Analysis* 29, 1994: 41–47.

[11] Kanis, J. A., Bergmann, J. F. "Full Consent May Bias Outcome of Trials." *BMJ* 307, 1993: 1497. Tobias, J. S., Souhami, R. L. "Fully Informed Consent Can Be Needlessly Cruel." *BMJ* 307, 1993: 1199–1201. Dahan, R., Caulin, C., Figea, L., Kanis, J. A., Caulin, F., Segrestaa, J. M. "Does Informed Consent Influence Therapeutic Outcome? A Clinical Trial of the Hypnotic Activity of Placebo in Patients Admitted to Hospital." *BMJ* 293, 1986: 363–4.

[12] Thung, P. J. "The Growth of Medical Knowledge: An Epistemological Exploration." In: Ten Have, H. Kimsma, G. K., Spicker, S. F., eds. *The Growth of Medical Knowledge* (Philosophy and Medicine, Vol. 36). Dordrecht: Kluwer 1990, 87–101.

[13] Bergmann, J. F., Ref. 10.

[14] Ballegaard, S., Pedersen, F., Pietersen, A., Nissen, V. H., Olsen, N. V., "Effects of Acupuncture in Moderate Stable Angina Pectoris: A Controlled Study." *J Intern Med* 227, 1990: 25–30.

[15] Ballegaard, S., Ref. 14.

[16] Bayley, C., "Our World Views (May Be) Incommensurable: Now What?" *J Med Phil* 20, 1995: 271–84. Veatch, R. M., Stempsey, W. E. "Incommensurability: Its Implications for the Patient/Physician Relation." *J Med Phil* 20, 1995: 253–69.

PIOTR MRÓZ

THE PHENOMENOLOGICAL ASPECTS OF MENTAL DISORDERS

The purpose of this paper is to present the principal features of an "existential phenomenology" represented by Jean-Paul Sartre and Maurice Merleau-Ponty. This philosophical trend (one cannot refer to it as a system with a complete, definite nature), which gained enormous popularity in the late forties and early fifties of this century, seems to have exerted a great influence on the emergence of a new kind of psychiatry. The "humanist outbreak"[1] (to use Sasz's term) is closely linked to the theory of the human project and "the desire to be".[2] We are of the opinion that the description as well as the analysis of human consciousness – propounded by Sartre and to some extent by Merleau-Ponty – helped to bring about substantial changes and transformations in so far as the attitude towards "deranged people" was concerned, closely following Brentano's "axiom" concerning the nature of conscious acts, Sartre and Merleau-Ponty promptly accepted the advanced forms of phenomenology taught and studied under Edmund Husserl. For new followers such as the French group, it appeared evident that acts of awareness must – by their nature – "have" an object.

This opened up a new vista for European philosophy. It stands to reason that consciousness was supposed to be the only universe of philosophical discourse which was valid and worthy of analysis. This previously unexplored link between something transcendent – an object out there – and a conscious subject (I, Ego or, to use the Sartrean and Pontian terminology, *pour-soi* and *en-soi*) seemed to be a remedy for all of the insoluble problems and issues that the European tradition did not manage to resolve. These were the problems in which metaphysics entangled itself and, under so heavy a pressure, could make no progress.[3] Thus, if the nature of consciousness was proclaimed to be "intentional" or "thetic", then the "world" along with the "I" were to be mutually interdependent. Moreover, a new quality (in a strict philosophical sense) was born: the intentional object. It was a kind of "merging" of both the world (standing out there in its massive passivity) and the thinking (aware) subject.

However, the teachings of Husserl were soon to lead not only to new implications, but to a definite rift between the old master and his disciples. Nevertheless, French and German phenomenology paid a great tribute to the author of "Ideas". It was widely proclaimed that he had been absolutely right

A.-T. Tymieniecka and Z. Zalewski (eds.), Analecta Husserliana LXIV, 75–84.

in discovering the authentic nature of awareness. The axiom admitting *the constitutive* role of each conscious act appeared to be quite revolutionary. But the rigorous stipulation that all acts could be reduced to one predictable and stable matrix, the so-called *Transcendental Ego*, evoked strong criticism on the part of Sartre and Merleau-Ponty.[4]

It was Sartre who, in his early study *La Transcendance de l'ego*, had openly rejected such an attitude towards consciousness. There is no phenomenological evidence – we find Sartre saying – either on the prereflective or the reflective level, for the existence of such an "inhabitant" of human consciousness. Moreover, the very nature of intentionality makes us regard human consciousness as "something" (or rather "nothing") that is initially empty. Sartre puts it bluntly: consciousness in the world is nothingness. It is certain, Sartre states categorically, that all forms of the so-called transcendental Ego would totally destroy the very idea of consciousness as the active, free agent building and forming the world. The latter – let us recall – is the result of conscious acts and intentions. Their role is to aim at something which is transcendent, that is, something that consciousness is aware of but cannot be equated with. At this moment we approach possible implications which might be offered to human psychiatry.

Let us now turn to the basic, fundamental modes of "working" or "acting" consciousness in the world. Husserl, and subsequently Sartre and Merleau-Ponty, described the mode of *perception* and *imagination* (not in the traditional, psychological sense, however) as the two ways of constituting the world. The first mode grasps reality in an ordinary, "normal" way. The pot I am perceiving at this moment appears against a real background which is composed of my study, my apartment, the street I live on, etc. All perceived qualities seem to be part and parcel of the infinite whole – *the hodological space*. The temporal aspect is of great importance here as well. The pot in question was being, is being and will be perceived in – respectively – the past, the present and probably the future. It would be a mere impossibility to grasp all its qualities at the same time in order to reach all the interconnections the pot is said to possess. In short, to bestow upon this rich and "open" item a definite sense and meaning. In a word, perception is a mode of working consciousness which step by step, act by act – as it were – tries to reveal the "true" face of a partly given reality. No matter how long the act of perception lasts, perceived objects will always hide some of their aspects while revealing others. In eidetic analysis, however, we acquire the knowledge that, in comparison with the consciousness performing such an act, objects are endowed with essence, nature, something "stable" and "constant". Being what they are, the objects of perception are "mine". They play various functions in my life and that of others. They are "seen as", "regarded as useful or useless", "judged", "valued", "disvalued", "taken into consideration", "chosen" or "totally discarded" and so on. In short, it is the perceptive consciousness that builds the world around us – the real world. In this mode of intending the world (and others), my judgements, systems of significations and intentions may be said to have a kind of intersubjective meaning. An object or still better a being-in-itself is perceived as something real, not only for myself but for other "subjects" as well. We – as a group of human beings, as free consciousness inhabiting one hodological space – may differ widely in thinking of (about), perceiving, passing judgements, social "pigeonholing" objects or persons. But they generally remain themselves, being interconnected with each other, being perceived on a wider plane (a "universal" plane). Merleau-Ponty sees in the perceptive mode of consciousness a kind of psychic norm, a yardstick by which we measure reality. Thus the norm is predictable in a way. Objects constituted in perception are hardly supposed to evoke any "surprise". However, the range of "surprise" may be wide. The pot in question is likely to fulfill many functions: my grandfather used it as an ashtray while my grandmother used to keep bills and scraps of paper in it. Nevertheless the pot remains the same forever. Getting "in touch" with the world out there, the perceptive consciousness gains a universally valid communicative dimension. An intentional object has a quality of *something-in-common.*[5] The norm – be it psychological or social – is retained and observed not only by my sole, unique, free awareness, but by that of others as well.

The main point of contiguity of existential phenomenology with humanist psychiatry rests, however, on the second aspect of consciousness. In order to see this link clearly we should now turn our attention to the other way of reaching the transcendent world. Before we enter the domain of the imaginary mode, we should devote some time to the most important part of existential analysis: the human project.

It should be remembered that according to Sartre and Merleau-Ponty, the nature of human consciousness can under no circumstances be reduced to anything beyond itself. Advanced forms of the Sartrean and Pontian analyses equate consciousness with human being itself. It is so different from the object it "intends" that any attempt at bringing them together would deprive human being of its most important qualities – spontaneity, freedom and translucency. We already know that consciousness – being conscious of the object – lacks the qualities which objects possess. In *Being and Nothingness*,

Sartre gives his famous definition of being-for-itself (man or consciousness). Since any single act of consciousness must by definition be a thetical operation, it stands to reason that all acts such as thinking, feeling, sensing or imagining are intentionally directed towards, or still better, intend, transcendent objects. The latter are given as subsisting outside our consciousness. It seems quite obvious – as the eidetic analysis should yield – that there must be an object (*out there*) and a "subject" (not in a traditional sense, however) *in here*. Sartre's point of departure – as B. Shul has pointed out – is the incontrovertible evidence of Descartes' *cogito*. This famous "I think therefore I am" is combined by Sartre with the concept of intentionality and the method of phenomenological reduction. Sartre underlines that the very formula *cogito ergo sum* immediately implies the following indisputable fact: I think, therefore there is an I (object) in the world. The I (object) which is in the world *is*, while the I (subject) which thinks *is not*. We quote: "Put into specifically Sartrean terms, there is a fullness of being (*en-soi*) which is the being of things, and a nothingnesss, being for itself (*pour-soi*), which is the being of consciousness".[6] The latter – let us add – is the very awareness of the being of whatever is its object, of the not being of whatever it reflects upon: the "for itself" is the negation of the "in itself". The next step, as it were, in reading the *cogito ergo sum* formula concentrates on drawing crucial consequences from this rift between consciousness and its multifarious objects. It is worth mentioning that, in spite of all evident and unquestionable affinities to the Husserlian version of phenomenology, Sartre proposed a richer, thus more humane or existential, understanding of all conscious activities.

These were not only supposed to organise (cognitively, axiologically or practically) the raw, opaque and unstructured reality, but also – so to speak – to "mark off" at all times this deep, dramatic difference between consciousness and the object it intends. The first analysis revealed to Sartre that, in contradistinction to the transcendent entities, consciousness appears to be a totally free, spontaneous and translucent force. Its "origin" – from an ontological point of view – shows that it is totally contingent, absurd, thus absolutely unjustified. In short: consciousness is a freedom that we are all condemned to for no reason whatsoever. In "asking about" the-being-in-itself, doubting in it, thinking of it, transcending it in various ways and processes, the *pour-soi* becomes none other than the negation of its antagonistic, hostile counterpart. Such contacts, or better yet, such dynamic relations with the being in itself (which is ontologically closed, defined and equipped with essence or nature) evoke uneasy if not traumatic experiences in every *pour-*

soi. The latter does take itself as a series of always open – till death comes – possibilities of the negation of the *en-soi*. Thus, the-being-for-itself (equated by Sartre with the absurd whirl, the vertigo of freedom) is a double nothingness. This truth (at the same time a final conclusion of the Sartrean reading of the *cogito ergo sum* formula) can be interpreted in two ways. First, the primary manifestations of consciousness may be grouped into such phenomena as "absences, gaps, missing parts in the total field of the being in itself" – to quote from Herbert Spiegelberg.[7] (These are referred to by Sartre as *negativities* which adequately describe each consciousness). Secondly, consciousness as "the being through which the nothing comes into the world"[8] must *ex definitione* be its own existence (so strongly opposed to the being in itself). This leads Sartre to his famous definition of the human reality (human being). It is, Sartre vehemently declares, an existence which is not what it is and is what it is not. Thus the difference between it and the being in itself which is only what it is (having no immanent power to "overcome", transcend itself) must emerge in a very intentional act, a choice of consciousness – the entity of free and spontaneous but essenceless "nature". For Sartre, existence (that is, a human being, the *pour-soi*) is invariably understood in terms of the consciousness of one's freedom in the world. As such it must always precede the essence – the meaning or sense of one's existence which the *pour-soi* may acquire through all its intentional acts or through its various projects.

As a contingent, absurd and "ecstatic being", consciousness experiences itself – prereflectively and reflectively – as totally unjustified. Let us note the fundamental feeling of being *de trop* so vividly described in Sartre's masterpiece *La Nausée*. Hence – Sartre states in his ontological analysis – when all the motivations of consciousness have been explained away, an ultimate moment of choice, irreducible to a single explanation, can be clearly discerned. It is the exercise (*aventure*) of our unlimited freedom from the given situation, and freedom towards the *projected* world, which has been identified by Sartre with the very structure of consciousness in the world. Since its primary trait is negation (consisting in transgressing its unstable, *nothingness-like Ego* intending the various spheres of being in itself) consciousness can never coincide with itself. To put it differently: as a spontaneous whirl it is constantly surging forth, tearing away from its ontological predicament, its "given facticity". So, consciousness is always restless, tortured and haunted by its ecstatic freedom, temporal diaspora and internal anxiety. In a well known description (tinged with typically Sartrean philosophical imagery), we read that consciousness hectically desires the density of the being-in-itself but, at the same time, the *pour-soi* would never agree to resign from or give up its status of a free, transparent and ever spontaneous force. This paradoxical and ambiguous project (*the original, ultimate choice*) is supposed to result in a simultaneous coincidence of essence ("that which is") and existence ("that which is not what it is and is what it is not"). Thus in all contacts with being-in-itself, transcendent reality makes being-for-itself actually aware of its ontological lack (*manque*). Sartre reaches the heart of the problem of human existence – its behaviour and norms – through his *ontologie radicale*. Considering the lack human beings are experiencing, they would like to attain the massiveness or ontological stability of being-in-itself, in other words, the essence that things are endowed with.

This dramatic, tense struggle to attain the above-mentioned qualities is called the general project. As Sartre puts it: "Human beings desire to become God".[9] In terms of his *ontologie radicale* it would be an impossible merging of being-for-itself with being-in-itself, so paradoxical that it is merely unthinkable. Nevertheless, all our intentions and acts can be reduced and explained by reference to the general project or choice. It goes without saying that the endeavour on the part of consciousness to obtain the qualities of being-in-itself is combined with the wish to retain the primordial character of human being: its freedom and spontaneity. In sum, human existence is nothing but the ontic paradox – an unbreachable gap between being and existence. A gap which is not to be transcended by the ordinary, *normal* mode of consciousness, perception.

It is high time that we concentrate on another mode, namely, imagination. In his famous studies[10] devoted to his general project, Sartre gives a vast but precise description of how "abnormal" beings and their works ("fixed intentions") can be looked at and explained. The message pervading these monographs is that the consciousness of a human being wants recuperation or reconciliation. ("Wants" is used here in its fundamental, basic sense.) To put it in other terms, one may say that we dramatically strive to obtain the nature we are deprived of: to be Peter or Mark who would be what they are, being free and unlimited consciousness at the same time. It sounds like a sheer impossibility. But for Sartre and Merleau-Ponty another mode of intentionality may in some cases heal this ontic gap between existence and being. In the case of Baudelaire and Jean Genet, Sartre shows how all forms of this attempt are realised. These are forms – let us add – which place an individual on the margin, whether socially, psychically, or to employ the terms of existential phenomenology, ontically. The maverick story of destination, fate, nature, freedom and essence has, of course, its beginning in early childhood – the most formative period in a man's life.

Hypersensitive beings prone to experiencing this ontic discrepancy wish to reach the merging of being and existence, freedom and essence. In the case of Baudelaire, it is the dream of an essence of being different from the others. Charles is slightly neurotic (naturally this term does not suffice to cover the range of his ontic strife). Deprived of a natural father, who left him "so long ago that one cannot remember",[11] Charles concentrates his dispersed, uncontrolled emotions on his mother. What he really wants is her "psyche and physis". To possess her totally would mean to make existence and being come together. Thus, the first phase is that of a merging with the existence he loves.[12] By incorporating another being-for-itself – Charles reasons – mine will be justified. As there is no definite cause of my coming here and now – no God to believe in, no history to continue, no parents who have any sense to exist for, I can accept the only possible choice: being different and showing it off among the others. This specific relation to being and existence (an attempt to make them meet) constitutes the concrete signs of his *maladie*.

All possible (traditional) explanations are doomed to failure, according to Sartre. "Psychic disorders" of all kinds – explains existential phenomenology – are, as a matter of fact, of ontic character. It is human consciousness which, in its endeavour to attain the impossible merging and recuperation in the act of free choice, takes up, so to speak, particular projects. These are supposed to fill the gap. In the case of Baudelaire the man, "deviations" such as narcissism, introvertism, homosexuality and acute neurosis are in fact examples of the general project. Sartre discards any psychobiological interpretation as completely invalid. Every human being (with consciousness, being-for-itself) is in a concrete situation in which he/she must choose. It is we who should be aware that we are nothing else but a form of *néant*. Nothingness, which is our freedom, is impossible to escape from. But man – it should be recalled – strives to be God, who is fully justified and endowed with nature, to be a homosexual who is free from homosexuality, to be Charles who is free from being Charles but is perceived as he wants to be perceived. In short, man strives to be free and to have essence at the same time. This dramatic tension between existence and being-to-be-never-reached is found most acutely in the mode of perception. I envy the ashtray on my desk because it is what it is. But this dramatic struggle is not to be won, underlines Sartre, in the normal, perception mode. The fight which is likely to be lost by Baudelaire the man (connected with others by intersubjective acts) has a good chance of being won by Baudelaire the poet. All deviations which are the signs of the general project are something *natural* at the moment when Charles switches over to the imaginary intention. The perception of reality is valid only in the context of its rejection. Things and human beings, relations and acts are wholly dependent on my imaginary attitude. They are given in one quick "glance" – they never threaten me, they are what they are because of my will. What is more, the sphere of imagination promises me the realization of the essence. By an act of creation, Charles – as any human being would wish – performs such an act, treating his poems as something ontologically completed, something reminiscent of a realized project, a being-for-itself (it is mine or still better, a part of myself) and a being-in-itself (each *fleur du mal* has an objective existence). Is it not a kind of God who is being revealed by particular stanzas of Baudelaire's poetic *oeuvre*? It stands to reason that the sphere of art fulfils yet another function. Since the fusion of existence and being is absolutely impossible in the mode of perception (that is in the world of our normalized feelings and experiences), its substitute might be possible in another mode of intentionality. The most important thesis is that what could be taken as the syndrome of a particular, mental disorder in the natural world of perception now becomes fully understandable in the imaginary world, or still better, *anti-world.*[13] The above assumption is not, however, the cornerstone of a reconstructed outline of mental disorders as one finds in existential phenomenology.

What is really important, to my mind, is the strong thesis that by the act of imaginary intention a specific object is born. It may be the convergence point of contradictory qualities. We do remember that the sphere of perception corresponds to reality, which is totally unlimited in its vastness and richness and is not to be embraced by a single, particular intention. "Things" (intentional objects) are what they are through being endowed with stable essences. We envy them their ontological status, being fully aware at the same time of the everlasting discrepancy, the rift between the existence that we are and the being that we are not. The mode of perception forces us to behave as others do, whereas the imaginary mode, along with the result of this intention – an imaginary object – takes us to yet another dimension. Traditional psychology or psychiatry would term it as something abnormal, "disordered", but for existential phenomenology this is the sign of the project – the primordial quality of our undefined and open nature. Through an act of imaginative creation, a given object so closely tied up with its creator – free, unlimited consciousness – is ontologically justified. For a moment the contact with the imagined world – the antithesis of perception, brings us closer to the dreamed of but never attainable fusion. This dramatic quest for the ontic solution of the human problem does not of course find its *finale*, its final resolution, as Ponty has it. Baudelaire and Genet (in the Sartrean vision the most extreme cases of the attempt of the realization of the general project) perceive reality as all "normal human beings" do. But their attitude (all traditionalists would name it in disused and worn-out terminology of mental disorders) is extremely different. Perception (the constitution of the intersubjective world) in their case leads to its immediate rejection. To negate the sphere of normality is to enter the road of freedom. The acute awareness of being and non-being at the same time must evoke socially disapproved behaviour. Genet's total depravity, viciousness, moral and social degeneration is nothing but an attempt at fulfilling the project. Born in the streets of Paris, the son of a homeless prostitute, not knowing his natural father, he soon experiences *le regard* of the others. They perceive him as an outcast, as a non-person who does not belong. (See how Baudelaire is perceived by the society: it is he who regards others as outcasts.) Not withstanding their ways of life, we can see here the road to freedom which we cannot escape. The nature which we lack – such is our *conditio humana* which we must try to *outlive*, transcend – appears to be unattainable, but in the absurdity of our existence there is no reason whatsoever not to take up a struggle for it. If Baudelaire and Genet (as any one of us) identify themselves with a kind of essence superimposed upon them by the others ("Look, it is Jean, the thief, debauch, renegade. Look, it is Charles..."), both of them do so in order to play the game with reality. The latter ceases to be the field of perception in which normal behaviours are supposed to take place. The field, the hodological space of Merleau-Ponty, is immediately transformed into the imaginary one. Genet, the incarnation of evil, the focus of all degeneration, psychic *éparpillation*, and Baudelaire, the neurotic in the full sense of this term, become for the aesthetic moment of experience freed from the weight and pressure of being, or rather existing as humans. The explication of their social and mental abnormality and maverick-like status refers to the ontic state of affairs, so to speak. The truth reveals itself on the horizon of subsequent *étapes* of human existence. Step by step, choice by choice, project by project, some of us (Genet and Baudelaire are epitomes of a human condition) become more and more aware of this dramatic rift: we and objects, we and things, we and others. Each intention on the part of our consciousness – either on the perceptual plane or on the imaginary one – admonishes us that we are totally absurd and unjustified phenomena. The dream of the never-to-be-given nature (essence) being-in-itself is the mainspring of our choices. The momentary identification with the *eidos* we sometimes receive from the outer world is no solution to the problem of existence at all. Such an identification is to be transcended by another free act and intention.

Thus no nature can be attained at all. The evident signs of a vast range of psychic *maladies* are only symptoms of becoming more conscious of a condition. Once again, Sartre and Merleau-Ponty underline the fact that the mode of perception allows for (to a certain, but limited extent) a psychiatric (medical) explanation of all abnormalities and disorders. The ontological explanation, however, makes us embrace the totality of our condition. First of all, we are – or still better – we gradually become aware of the insurmountable gap between "things" and "others". In a word, the transcendent world out there, and our free, unjustified consciousness: we are doomed to exist. Secondly, we are dramatically conscious of the lack of essence we so strongly desire to attain, but under one condition – to be *ens causa sui* – God. But that wish is never to be fulfilled. The road to the general project is a source of all our sufferings and tremblings. The latter, as existential phenomenology underlines, must be interpreted and explained (if not healed) in terms of the general project. In this respect, what is termed existential phenomenology makes a significant contribution to resolving the everlasting mystery of existing as a human being.

Jagellonian University

NOTES

[1] This term is often used in Laing's and Sasz's works; see for example, *Psychiatry, Anti-Psychiatry.*

[2] See Jean-Paul Sartre, *Being and Nothingness,* New York, 1981, p. 542.

[3] See Maurice Merleau-Ponty, *Sens et non-sens*, Paris, 1968, p. 42.

[4] See Jean-Paul Sartre, *Imagination*, Paris, 1980, pp. 150–152.

[5] See Jean-Paul Sartre, *Being...*, op. cit., p. 390.

[6] Ibid., p. 90.

[7] Herbert Spiegelberg, *The Phenomenological Movement,* The Hague, 1969, chapter on Sartre.

[8] Ibid.

[9] Jean-Paul Sartre, *Being...*, op. cit., p. 490.

[10] We refer to the studies of Genet and Baudelaire as interpreted by Sartrean existential psychoanalysis.

[11] This is the fragment of a letter written to his mother from the boarding school his stepfather had sent him to.

[12] See Jean-Paul Sartre, *Baudelaire*, op. cit., p. 50.

[13] We refer here to the famous conception formulated by Sartre in his *L'Imaginaire*.

JAN HARTMAN

ILLNESS AS EVIL AND ILLNESS AS GOOD

It is bad to be ill. Illness is pain – a cardinal form of evil, at par with weakness, sin, wrongdoing, and disgrace. Among all the forms of evil it is perhaps illness which most directly and urgently calls for action. This action – treatment – is good, and as such it confers its own value upon everything that makes it possible. In the light of this understanding of good, illness itself acquires a secondary, indirect, yet nonetheless real value; in other words illness is also good. In the language of classical metaphysics, factors like this are referred to as "transcendental good": a type of good inherent "by default" in even the most wretched constituents of our essentially good (as it is God's creation) world. I will not, however, attempt to claim these metaphysical heights. On the contrary, I will confine myself to a pragmatic methodological perspective.

My paper, as the title suggests, will be devoted to the ambiguity apparent in reactions to and opinions about illness. I will present an axiological dialectic whose dynamism provides grounds for the dynamism of the human practice of treating diseases. Similarly, stimulating ambiguities and dialectics are encountered in the area of crime and its prevention as well as in the complex relationship between individual egos and state institutions that enforce action orientated towards a common goal. In all these cases, an initial evil becomes a condition for the emergence of a highly developed product of culture, be it the science of medicine and health care, or the legal system, or democratic institutions.

I will not attempt to present a systematic analysis of the theoretical and practical problems in the area described in the title. All I would like to do is to show how the aforementioned axiological ambiguity comes to the fore in different, more or less theoretical notions of illness and treatment, as well as in popular beliefs. In order to accept those points of view which see "illness as good", a departure from the classical concept of good is needed.

Namely, I am asking you to put aside that concept located at the crossroads of general metaphysics and practical philosophy and defined in the following way: *good is whatever is the direct or indirect goal of action, in particular of human action.* The concept of good that I have in mind is perhaps equally classical but, somehow mysteriously, competing with the former, namely: *good is whatever contains, creates, or enhances an enriching meaning – which leads towards perfection.*

A.-T. Tymieniecka and Z. Zalewski (eds.), Analecta Husserliana LXIV, 85–90.

There are numerous phrases that describe the decline from perfect health (to death) and accentuate different kinds and aspects of such a decline; some of them express the philosophy of the speaker as well as describing the illness.

Here is a list of some words and phrases that describe this decline, together with explanations of their meaning: 1) "Pain" – bad, negative physical or psychological sensation. 2) "Weakness" – feebleness, lack of vital powers. 3) "Poor health" – a general condition of the body, being prone to disease. 4) "Ailment" – a slight but persistent malfunction of an organ or organs. 5) "Disease" – an impairment of normal physiological function, a pathological change: we refer to this situation by saying simply: "I am ill". 6) "Hurt, injury, wound" – a sudden and serious malfunction, a break in the structure and violation of the integrity of the organs, even persistent damage to a part of the body. 7) "Disability" – a persistent handicap, malfunction or lack of an essential organ(s), seriously affecting one's everyday life; either inborn or resulting from a disease or accident. 8) "Monstrosity" – a marked structural deformity of an organ. 9) "Symptom" – any unusual sensation or change in bodily functions ("something's happened to me"). 10) "Syndrome" – a theoretical medical notion describing a well-defined set of phenomena and processes conceived of as bad for an organism.

We have some of these bad things in mind when we utter the word "illness" or "disease" in various contexts. The relevant semantic field is further broadened by different beliefs and ideas of what really happens inside us when we are ill – in any of the above meanings. These beliefs and ideas can be quite simple, or can amount to entire medical theories, but all of them are grounded in general beliefs about what illness is in general and what its particular types consist in. This popular ideology and typology of illness seems to be worth analysing; not the least interesting is the fact that the categories of diseases are defined by certain "cardinal diseases" and "diseases as cultural symbols": leprosy and syphilis, for instance.

1) The most primitive belief is, of course, the mythological idea of illness as "being possessed by evil forces". The idea of mental disease as "possession by the devil" and of infectious disease as "bad air" both conform to this paradigm, as well as the explanation of illness as a punishment sent by God. These all are versions of the myth of disease as the intervention of an external super-human power, an intervention that is caused either by an infringement of a taboo, or as a result of sin. Illness in this sense is always accompanied by moral taint, a disgrace on the ill. Its symbols are the leper and the lunatic.

2) Another classical view of illness, or at least of one of its essential types, is the idea of disequilibrium, of upsetting the balance of certain regulative components of the body – blood, phlegm, yellow bile, and black bile, for instance. A paradigmatic disease in this sense is sickness or intoxication. The notion of illness as a disequilibrium and disharmony turned out to be exceptionally fruitful for the science of medicine. It was this concept that enabled the change from treatment in the form of exorcism to treatment in the form of "detoxification" by applying a specially prepared substance: a remedy. This concept paved the way for modern pharmacology.

3) A medical topos of "bad blood" links the early modern concept of illness as an alchemical violation of homeostasis with the concept of illness as partial death: decay, rot, or gangrene. The prominent diseases here are "the vapours" and syphilis. Treatment consists mostly in "purification" of the body: surgical removal of decayed parts, amputations, cutting ulcers open, letting blood. From the desire to remove all rotten bodily matter arose the most military of all medical disciplines – modern surgery. A sort of medical conservatism is also grounded in the same concept, namely, an approach that considers many diseases as relatively natural phenomena, similar to aging, and thus, unworthy of intensive treatment.

4) Yet another modern view of pathology – the mechanistic one – is symbolised by paralysis. Disease is a fault in the bodily mechanism – a malfunction. Something has "gotten stuck", a cyclic process has been interrupted, a circuit has been cut, a vein has been blocked. The idea has proven to be extremely important for scientific medicine, since it emphasises the phenomena of conductivity, metabolism, and circulation. The search for a motor of the bodily mechanism directed the efforts of medicine toward studying the functions of the heart. This spiritual supremacy of cardiology is still prevalent in contemporary medicine – and certainly the fact that cardiac arrest is the most common cause of death is not the sole reason for it.

5) The most awesome disease of all – cancer – symbolises another type of illness: illness consisting in turning the vital powers of the body against itself (illness as a tumour, growth, monstrosity). In this category illness can be described as self-aggression, self-destruction, and anarchy in an organism. Neoplasms, acromegalies, and allergies – understood as auto-aggression of the body – are paradigm cases of diseases that support the idea that the ultimate cause of illness is the patient himself.

Having surveyed, though all too briefly, the morphology of the evil of illness, keeping in mind the whole network of various "bad associations" aroused by the word "illness", we are now in a position to recognise their mirror images in the concepts and ideas through which illness manifests itself as a non-evil, and indeed, as something positive and in a sense good. The existence of this positive reflex is inevitable for pragmatic reasons. Since illness is the necessary prerequisite for the existence of therapy and medicine, it must be tamed, accepted and explained by those who live with disease and those who live off it – by patients and doctors. Like every inevitable thing, illness must be presented as something meaningful and valuable in the broad perspective of the human condition.

All philosophical theories of illness, maybe except pessimistically existentialist philosophies, strive to discover a sense in illness. The Christian concept of illness is ambiguous in this respect. Learned Christianity, on the one hand, teaches that illness is a trial through which God "tests" our faith. Illness is, moreover, a *memento mori* – a figure of death, and a path to death. It therefore has a significant religious dimension; it is one of the ways to God, an opportunity for penance for the sinner, an opportunity for conversion for the infidel, and generally an opportunity for diverting thoughts from everyday mundane tasks to matters connected with salvation of the soul. On the other hand, however, popular Christianity tends to see illness mostly as something bad that God allows to exist – therefore it has to be borne, and accepted. Sometimes illness is even seen as God's just punishment, implying that whoever is ill must have deserved it.

The classical Greek concept of illness is most manifest in the stoic heritage. In this tradition, illness appears to be "insignificant". The virtuous man ignores it, or, more precisely, accepts it as a fact of life towards which his attitude should be one of indifference. Illness does not encroach onto the sphere of his spiritual autonomy. Undergoing treatment is perhaps advisable, but only in so far as it is advisable to exploit all favourable opportunities – but without excessive effort to bring them about. From this point of view illness is not so much an evil as an obstruction.

Doctors of antiquity knew how to treat many diseases quite well, and scientific medicine existed in the late Middle Ages too (mostly thanks to Avicenna), but it was not until modern times that the idea of an active fight against disease came into being. We owe its birth to the modern spirit in philosophy. But not entirely and not everywhere. For the modern spirit in philosophy was far from homogeneous, and could be divided – provisionally – into two intellectual trends: a naturalist, positivist one and – in contradistinction – a spiritualist and romantic one. It was the former one – evolving from 16th-century medical alchemy to positivist somatism, symbolised by the cult of anatomy and vivisection, that inspired the search for active methods of remedial treatment. Illness was becoming more and more axiologically neutral: a natural, though unwelcome, state of an organism, as natural as all other diversions from norms which abound in nature. From this perspective health is a norm, whereas illness is an abnormality. On the other hand, romantic modernity is responsible for the "aesthete" concept of illness (we will come back to it later), as well as for the revival of spiritualist magical practices. These were dominant in the Renaissance, and today they resound in several – sometimes moving, despite their naivety – descriptions of efforts of a seriously ill person to regain health as if they were heroic battles with a cunning enemy who nevertheless can always be repelled with spiritual strength and an indestructible will to live.

Philosophies, or rather ideologies, of illness influenced not only general concepts of medicine, but also the practical attitudes of doctors as well. A conservative doctor with Christian or classical attitudes is by no means quick to opt for active and aggressive treatment. Illness for him is a natural state and most diseases, as he is always ready to remind us, are self-curable. Moreover, every age has its diseases, and the biological clock cannot be outmaneuvered. When the time comes for dying, for arteriosclerosis, a stroke, or cancer, the patient should be allowed to die quietly. Intensive treatment is only justified in cases where minor diseases turn into serious ones, or when a dangerous disease occurs "unjustly" – the patient is too young and "too healthy" for it. It should be said that such traditionalists among doctors are rather rare today. The active type is dominating – and not only because modern ethics demands that the doctor *do everything possible* for the patient, but also due to the fact that in a contemporary free-market society, doctors – like everyone else (including philosophers!) – tend to sell their services as broadly as possible. An active doctor, thus, treats you even when you are as fit as a fiddle; for example, he/she may recommend prophylactic treatment, or a diet and a healthy way of life; or he/she may prescribe vitamins, cosmetics, and Prozac (if you happen to live in the US). If you feel just a bit under the weather – and have made the mistake of telling him/her about it – he/she will come up with the most sophisticated treatment. Nobody is allowed to be ill in his/her presence – let alone die. In the world of active medicine, the boundaries between illness and health, prevention and therapy, serious disease and hypochondria, have been obliterated. This is perhaps why "allergy" seems to be the favourite of contemporary medicine.

As we have seen, in the complex history of medical beliefs and ideas, illness was looked upon not only as a definite evil but also as a manifestation of the natural (and supernatural) order of things, as something irrelevant and trivial but also as something significant and important. Human activities oriented towards the fight against diseases were, and still are, being amplified by the forces inherent in tensions between various ideas, concepts, beliefs, and opinions connected with diseases and treating them. It is not worth talking too much about the fact that it is bad to be ill. It seems, however, worthwhile to point to another extreme in the whole range of opinions that have ever occurred in the philosophy of illness. This is the "aesthete" approach that has already been mentioned, the idea that illness has certain specific inherent moral and aesthetic values.

The aesthetics of illness is grounded in a *prima facie* justified belief that the socio-psychological type of an individual is partially determined by his/her physical appearance: the way he or she looks, his/her age, the condition of the body. The unique personal quality of an individual is thus often co-determined by certain flaws, imperfections, and disabilities. It is hard to be a noble Nestor without being aged; a justified user of a walking stick must suffer from at least a slight impediment of movement, etc. The Romantic movement valued the aesthetic aspect of illness to an extent that provoked some to *pretend* to be ill (e.g., by using pale powders) and even to engender it artificially (e.g., by drinking acetic acid). Romantic ideals emphasized connections between the hapless life of an artist and the content and quality of his creation. A persistent and unbearable disease often figured prominently among the necessary calamities of a respectable artist. Thus illness was perceived as both a manifestation of and the reason for the artist's sublime sensitivity. Moreover, it guaranteed the truth of his creation. Great "patients" in the history of culture include Marcus Aurelius, Pascal, Nietzsche, and Proust. Among diseases allegedly connected with high spiritual values, tuberculosis and various heart problems can be mentioned, along with – even more significantly – all diseases perceived as specific to upper social classes, such as haemophilia, migraine, and anemia. The commoners, who were also not incapable of appreciating the value of illness, had to content themselves with illnesses which had "socially initiating" value and symbolised transition to the next class in a hierarchy defined by age (generation). Such "illness of age" included first the diseases of childhood, then diseases of middle-age – going bald and putting on weight (if we may count these as quasi-diseases), and finally diseases of old age.

There can be no doubt that a world without diseases would not be a human world; neither would a world without senility and dying. This is ultimately why doctors will always remain suspect individuals who tend to intervene in the natural order of things in this "best of all possible worlds".

Wroctaw

MARIA GOŁASZEWSKA

THE AESTHETICS OF MEDICINE:
An Essay on the Borderline of Aesthetics, Art History and the History of Medicine

First, I wish to explain the concepts included in the title of this essay. The word "medicine" comprises several meanings. In a large sense, "medicine" is a widespread phenomenon in the anthroposphere, a phenomenon whose essential moment is caring for human health (though humans also care for the well-being of animals and plants). Yet this generalized definition, which includes everything that lives and is treated by humans, is broad. It includes the treatment of animals, called veterinary medicine, and the treatment of plants, called phytopathology. Nevertheless, this broad understanding of medicine is not devoid of sense. It includes common elements like the human being's care for health, concepts of illness, and the Hippocratic oath confirming the doctor's care for the patient's health, and the patient's life as the highest moral imperative – the principle *primum non nocere*. Obviously, these principles are applied in different ways to humans, animals and plants. There also exist complicated or even terrible connections between medicine and the world of animals and plants (e.g., medical experiments on animals, herbal drugs).

Our direct object of interest is medicine in the strict sense, i.e., everything that is connected with the treatment of humans. Medicine is divided into a knowledge of human health and illness, and the ability to prevent illnesses, to cure the ill. We shall take into consideration both the historical and the practical aspects of medicine. Yet we shall be interested not in medicine itself, but in its reflection in the mirror of art – in how the image of medicine is artistically created. Indirectly, we shall be dealing with social consciousnesses, with opinions about medicine, and with the non-physician's views on health services.

The second concept mentioned in the title is "aesthetics". The object of aesthetic studies is an "aesthetic situation", a situation including the artist, the work of art, the recipient, and the aesthetic value. The relations of systematic interdependence between these elements are exceptionally important here.

Another domain of aesthetic studies is the beauty of actual reality: the aesthetics of reality deals with aesthetic values characteristic of nature,

A.-T. Tymieniecka and Z. Zalewski (eds.), Analecta Husserliana LXIV, 91–101.

events, human history. "Aesthetization" is a concept semantically and semiotically derived from aesthetics. It denotes the relations of actual reality with beauty: humans perceive facts, phenomena and events as simply existing, but they also strive to structure them, to introduce some order. Sometimes a human being builds certain wholes that are influenced by artistic or para-artistic structures. He or she creates, as it were, the images of works of art. What seems amorphic, incomprehensible and incidental in the pre-reflexive perception, gets submitted to the operation of "aesthetization". Artistic structures are applied for description, analysis and interpretation of that which is real (e.g., someone's life story as a novel).

Both of these moments – references to art and to the beauty of reality – are introduced in our considerations, endowing the "aesthetics of medicine" with a spatial dimension (apart from the already mentioned temporal dimension). Historical order prompts us to start the consideration from the ancient symbols, myths, legends, fairy tales, and, gradually, to get us to the modern means of artistic expression, such as novels, movies or television.

The history of the aesthetization of medicine cannot be reconstructed faithfully, but its essential features are perceptible to a certain extent. Historical records point to its mythical-magical roots. Both the illness and the ability to bring relief to the sick are associated, at their roots, with notions of a religious nature – first with animism, and later with the developed polytheism. Shamans endowed with the special powers of communing with spirits, simultaneously create full fantastic images, pictures of the world of joy, health, suffering, illness and relief. They perform operations that are half medical and half magical – they perform skull trepanations, evoke hypnotic trances, fumigate, recite incantations. In the prehistoric "ambulatory", a doctor-priest arranged a mystical ritual-performance. He created a specific medical miracle play in which he introduced faith, apart from the medical procedures, in a direct intervention of the world of spirits into the world of human beings.

One of the motives of Ancient Greek medicine was the serpent symbol, a symbol which has survived till today. It came from the god patron of medicine, Asclepios. It was Asclepios who ascribed to the serpent a specific curing power. His godly daughter, Hygea, was represented in Greek art as a young girl with a serpent wreathed around her arm and a medical goblet in her hand.

The medicine of Ancient Greece gradually acquired independence as a science and art of curing. Philosophers like Democritus or Aristotle (who came from a medical family) occupied themselves with curing people. Hippocrates, the author of medical treatises, from the Island of Cos, is recognized as the father of medicine. He already knew the systematics of diseases as well as the concepts of diagnosis and prognosis. His hypothesis was that diseases are caused by a defective composition of the organism's fluids such as blood, mucus, bile. His most important work, however, was the creation of the foundations of medical ethics. Here, his practical imagination produced a specific, artistic-like vision of speculative medicine. It is a theory of curing as a spiritual activity in which the specific people endowed with healing powers, great knowledge and high morals are crucially important. The art of curing can be mastered only by exceptional people who acquire medical acumen in a transcendental way. This orientation was supported particularly in the center of Epidaurus, the city that was simultaneously a medical center and the center of Asclepios worship. (Theater was used there as one of the forms of therapy).

The mythologization of medicine was intensified in the Middle Ages. It must be remembered that the ancient medicine made considerable progress in those times. As early as the third century B.C., the rationalistically and empirically oriented physicians like Herophilos Erasistratos dealt with anatomy, and performed autopsies or biotomies. In ancient Rome, physicians continuing with Greek medicine distinguished between the domains of sitology, balneology, surgery, gynecology, and obstetrics. The knowledge of drugs developed considerably. In the Middle Ages medicine was practiced, but exorcisms were much more important than the medical treatment. Illness or health were issues for prayer, faith, good spirits and guardian angels. They were the manifestations of the evil activity of the devil. Yet in the late Medieval period, the first faculties of medicine were created at the universities of Bologna, Paris, Montpelier, Prague and Cracow.

A good example of a disease being treated as a diabolical case of the spirit's impurity can be found in the fragment of Vit Stwosz's altar in St. Mary's Church in Cracow. The bas-relief *Christ in Limbo* presents the figures of three devils. "The faces of the devils bear the signs of rubella and syphilis reproduced faithfully and with a precision of detail allowing today for a medical diagnosis", writes an art historian analyzing Vit Stwosz's work.

The closer we come to the modern days, the clearer the picture of medicine becomes. The father of modern medicine, Paracelsus, recognized the art of curing as a great achievement of natural science. He broke with the views of past authorities whom he accused of superstition. In place of the complicated medieval recipes and magical potions, he introduced simpler cures.

Another criticism of medicine – this time the criticism is revealed in the work of art itself – comes from the 17th century: Molière wrote three theatrical plays (*L'Amour médecin, Le Médecin malgré lui, Le Malade imaginaire*) in which he accused 17th-century French medical practice of petrification and ignorance, of swindle and greed. As he himself suffered from consumption, Molière's death had, as it were, symbolic and theatrical repercussions; he died after the fourth performance of *Le Malade imaginaire* in which he played the major role.

The Polish drama *Nie-Boska komedia* (The Non-Divine Comedy) by Z. Krasinski dates from the period of Romanticism. The motive of illness occurs several times here. Being informed of his wife's mental disease, the poetically minded husband delivers a solemn monologue of despair. In response, he hears "a voice from somewhere" stating ironically: "You are composing a drama". There is also a scene describing a doctor who visits an over-sensitive blind boy, a child getting into a poetic trance. We also find in this play a shocking scene from the "asylum", a scene with a fairly detailed description of the genesis of illness. In this scene, we are moved by the "voices" which the mentally deranged woman hears. We can almost hear her death.

In the sociological realm, we can speak of "fashionable illnesses". They occur in a given period, in a definite area, and for no particular medical reasons. For social or snobbish reasons, it is often fashionable to fall ill and suffer. One of the fashionable illnesses is the "refined ladies' migraine" which caused the suffering of 19th-century ladies. The illness, or, rather, the artistic descriptions of this illness, portrayed women as "weak creatures" requiring care, understanding and tenderness. Interestingly, due to auto-suggestion mechanisms, the symptoms of this illness may really occur (as with hypochondria and hysteria, as well as with migraine).

Drastically more serious are the dangers of epidemics. The Middle Ages knew great epidemic calamities such as cholera, bubonic fever, leprosy, and group hysteria. Numerous paintings as well as literary descriptions are based on the motives of these calamities. We could hardly describe these phenomena as "fashions". Rather, they resemble common madness, in the background of which one can see inevitable death, usually as a punishment for sins.

The later "diseases of the century" became more and more rational, explicable. Yet an individual or a group has always been asking the question: "Why me?", "Why us?", or, the questions more familiar to the people indirectly involved with the art of medicine: "Are diseases controlled by chance, or by medical negligence and insufficient knowledge?"

The disease of the 19th century and of the beginnings of the 20th century is tuberculosis. It is suggestively described in the T. Mann's novel *Der Zaube-*

rberg, or in the poetry of Liebert who, basing his art on his own experience, created his lyrics in response to his illness. The poet suffered from tuberculosis, and he died before appropriate medical drugs were invented. In his four lyrical poems, Liebert speaks of the real aspects of his illness: the changes of body temperature, the helplessness of the doctors, the deterioration of lung tissues:

How lightly, lively, mercury jumps
and the column in the thermometer grows –
thirty-seven, point, five degrees!
Oh, it starts so innocently!
Just the breathing growing faster,
and dark patches below the eyes....[1]

Mortally ill, Liebert consciously structures his imminent death in a resigned and acceptant mood:

Through the fragrance of pines
They will go – each in its direction
The body – into the soil of June
The soul – to the green mountains.[2]

One may say that this poetic, ironically joyful acceptance of the inevitable is of some therapeutic value; internal peace protects the author from despair, mental depression, neurasthenia.

The 20th century is characterized by a weakening of the nervous system, hence neuroses, hysteria, neurasthenia, psychoses. Mental discomforts like stress or depression are included in the psychiatric registers of illnesses. Mental illnesses have their reflection – like in a mirror – in works of art: J. Krzyszton's autobiographical novel *Madness*, Elias Canetti's *Auto da Fe.*

The second half of the 20th century is characterized by the common interest in handicapped persons. Widespread social actions and movements have developed around this painful problem. Numerous movies, mostly documentaries, are made on this subject. Finally, there is AIDS – the disease terrifying people of our times. Here, also, numerous films assume the form of reports, para-reports, fabulated reports, etc.

Medicine in the mirror of art, indeed an extensive subject, shall be discussed here in two examples from the 20th century: *The Plague* by A. Camus, and *The Cancer Ward* by A. Solzhenitsyn. These are real masterpieces of their genre; with a well developed stratum of general ideas, they touch the deeper layers of social life.

The Plague is a novel about a doctor who, by giving up his moral and philosophical motivations, fulfills his professional obligation with full devotion and competence. "It is as it is", is the creed of Dr. Rieux from Oran. This creed resembles the immemorial "I AM who I AM". Both these sentences are tautologies, and as such they do not require arguments for their justification. A physician is simply a physician.

Obviously, a complete analysis and aesthetic interpretation of Camus' work extends beyond the frame of this essay. The book is a faithful description of the plague in Oran. Its action evolves from the plague's first symptoms, and – through the reactions of physicians, patients, their families and friends – to the horrifying later symptoms, to the death of the city dwellers and the collapse of the whole social environment.

When the epidemic becomes something obvious and medically identified, we are taken to a church to hear the sermon which attempts to find the meaning of what has just happened. And this meaning is of moral and transcendent dimensions. Thus, we learn that the plague is a punishment for sins. The priest relentlessly recalls the scared texts, he relentlessly scathes the parishioners with his gloomy preachings: "Brothers mine, a calamity has reached us, brothers mine, you have deserved it". Yet, at the same time, Camus reveals the hazards of this kind of attempt to save the meaning of life through summoning the God who is just but cruel, benevolent but vexed. And here we see – shocking in its medical realism and precision – the description of death: "In that moment a child, as if stung in the stomach, bent again with a shrill moan. And so he remained, bent, for long seconds, shivering and trembling convulsively, as if his fragile skeleton gave way under the furious wind of the plague, and cracked under the repeated blasts of fever". This description is long and cruelly precise. And when the boy dies, the doctor accuses the priest: "Well, this one, at least, was innocent, you know it well!"

What seems most important in Camus' novel is the clear and indisputable representation of the heroic figure of the doctor who is not prompted by any philosophical or religious motivation, but who gets into the height of medical service motivated only by "hope against all hope". He is faithful to the Hippocratic oath.

The Cancer Ward by A. Solzhenitsyn is a novel which we can unhesitatingly recognize as the artistic mirror of medicine. Through the objective picture of medicine (the doctors, the hospital, the patients, the therapeutic means) we get to know the relations between doctors and patients. We can also learn about the socio-political environment as well as the sketches of psychological portraits (since "a doctor is also a man"). The novel is an extensively realistic study, and all the thematic threads it spins are meaningful and important.

But how do the novel's characters perceive the progress of knowledge and the art of medicine? How do they judge the medical system? There is a voice of hope: "Obviously, it is painful to think that in three or four years in this century of all inventions they are sure to invent an effective cure for melanoblastoma". Yet, at the same time, there occurs a sign of skepticism: "Pooh, the sacred science! If they were such certain truths, they would not be exploded every ten years. And what am I to believe in? In your injections?"

Yet, overall, the novel represents an attitude of high esteem for medicine. It shows several excellent physicians of high professional ethics and the efficient work of the great mechanism of health service. The defects, which may surely occur everywhere, by no means decrease the worth of the system. What is most important, however, is the fact that the doctors themselves are aware of their shortcomings, and that they are self-critical. And so, one of the eminent doctors reflects: "If these doctors were fewer; if each of them was an expert and not simply an employee receiving a salary; if they were not thirty patients for each of them; if their minds were not marked by the thought of what it would be best to write in the case history".

One of the stronger artistic tricks the novel uses is the creation of a paradoxical situation in which a villain behaves like a hero, and the predator behaves like the victim. A mean patient, a party member "who is always right", angered by his uncomfortable situation of being a cancerous patient, tells his wife: "And these doctors in general are a hideous bunch! What right do they have to speak of production achievements if they work on a live man?! I don't know about you, but I hate doctors! – Yes, you are right, I have already told them here what I thought!"

Solzhenitsyn's novel strikes an aesthetically sensitive reader with the subtle composition of the author's medical knowledge. His erudition in this field lets him create the situations so attractively that the novel seems like a sensational plot with sociological analyses and romantic threads. At the same time it praises the fortitude of the doctors who – with unfailing tenacity – want to treat and cure, and the patients who – having enough of a will to live – overcome the disease crises and thus cooperate with the medical art of their protectors.

The most modern carrier of medical and para-medical phenomena is television. The TV series show the panorama of artistic approaches most fully, from the deeply moving tragedies of incurable illnesses to the comic conventions. The American series *M.A.S.H.* is a popular medical comedy, a successful attempt at making us laugh at scenes taking place in an army field-hospital. A surgeon, having helped an Ethiopian war casualty, experiences a moving moment: the saved patient, in an outburst of gratitude, kisses the surgeon's hand that was so helpful to him. He reacts: "It was the highest price I have ever been paid in my medical career".

The atmosphere of the American series *Chicago Hope* is slightly different. Here we can detect suspense, and sense the danger. When a youngster on the margins of society shoots at the hospital personnel with no rational motive, he is wounded and becomes a patient of those at whom he shot. The surgeon, though ill-disposed towards the assailant, has to operate on him. The patient dies during the extremely complicated brain surgery. The moral tragedy starts when the doctor questions himself: "Maybe the operation was not successful because I did not want the delinquent patient to survive? In spite of everything, I should have thought only about saving human life".

There are also movies expressing the tragic helplessness of medical service. One of them is a film about a doctor willing to provide medical help to Africans suffering from drought and famine (an American film, *We Are Children No More*).

The American TV series *Dr. Quinn* tells the story of a 19th-century medical pioneer, a young female doctor who is fighting for equal rights for women in the medical profession.

The series evokes our friendly feelings. The main character is shown as a noble person, courageous and slightly sentimental, but coping well with purely professional difficulties.

The aesthetization of the medical personality consists of a specific fabularization of the doctor's profession. On the basis of sometimes unclear facts saturated with emotions – facts having their own dynamics of rapid changes, like surgeries or spectacular disease symptoms – social images are created. These images involve medicine in general, but focus on the physician's work, behaviour, and habits.

Stories about known physicians sometimes turn into legends. Sometimes they are created by doctors writing about themselves. Axel Munthe and his autobiographical novel *Book of St. Michele*, L. Hirszfeld's autobiography *History of One Life*, and *Pages from Dr. Twardy's Diary* by J. Aleksandrowicz are some examples. Many more autobiographical stories have been published, e.g. Z. Flukowski's *The Years of Emergency Service*, or the collective work *Doctors' Memoirs*.

Medical professionalism is, as it seems, the most rational, functional option. In fact, it belongs to the line of technocratic thinking. The task of writing a work of fiction based on an actual life story of a physician fulfilling his or her duties seems easy. It may be easy, indeed, if the physician fulfills the medical duties perfectly: the perfect physician is a person who has knowledge and talent, who is always equipped with all the materials he or she needs. The essence of the "perfect physician" is the compound of education and medical knowledge. It is the *condition sine qua non* of a doctor: necessary though not sufficient. Though in the novels on medical topics there sometimes appear figures of "incompetent leeches" employed in hospitals, they got their jobs there due to unprofessional dealings (e.g. backstairs influence, bribery), and are kept at the physicians' jobs working as assistants, not as independent doctors.

Being "perfect" doesn't mean "behaving like a machine". A professional doctor working like precise cog in a great hospital machine does not gain social approval: he or she is sometimes accused of being "formal", "stiff". A doctor manifesting his or her "routine" or "automatized treatment of the patient" may not be liked by the patients.

The most optimistic though maybe not fully reliable concept is that of a physician as a person fulfilling a high social mission. The indubitable assumption here is that humanity is the highest value (no definite individual properties are taken into consideration). It is Albert Schweitzer, a philosopher and theologian, a missionary and musician of German-French origin, who becomes such an ideal representative of the medical profession, a legendary doctor. This highly versatile man characterized by high morality is also the author of the memoirs *Among the Blacks at the Equator*. Schweitzer professes the pantheistic ethics of respecting life. He claims that a physician should bear the unrestricted responsibility for everything that lives. Probably, a reconstruction of the very facts from Schweitzer's life as well as opinions on his position as a doctor and his practical activity could adequately exemplify our conception of the aesthetization of life, i.e., the structuralization of history according to artistic models.

In the synthetic approach, the problems of medical aesthetics may be discussed only selectively, since the phenomenon in question is highly complicated and too extensive to let us construct a systematic whole in a short paper. Therefore, we shall just mention the points that seem to be the most meaningful and important:

Practice. Practicing the medical arts is done in two ways: as an expression of personality, and as a purely professional activity understood in a technocratic way. Physicians of the first type get deeply involved in the treatment processes. They try to acquire comprehensive knowledge about the patients and to establish a direct, sometimes even slightly emotional contact with them. In the second case, the attitude of objectivity and submission to the established treatment dominates. The second case is oriented towards a narrow specialization surrounded by medical technological equipment.

"Fascination with a disease". This is a specific phenomenon that occurs relatively often. Physicians manifest such a great interest in a given illness or a given patient that dealing with these matters not only takes too much of their time and energy, but it also gives them a specific satisfaction consisting mostly in the feeling that they are dealing with something important which they can directly influence. Thus, it is a fascination of a creative nature. The patient's fascination with his or her own illness has a pejorative character: it is an experience of destruction taking place within the sick person.

"Searching for the meaning of illness". This is the way to diminish the suffering. It occurs frequently, especially in common-sense, current thinking. Physicians generally view illness as a matter of chance, heritage or negligence. In the social, common opinion, illness is sometimes understood as an inevitable fate, a punishment for sins, or a specific religious experience: a sublime, spiritually fortifying experience. All these attitudes find their reflections in art.

"Illness as weakening". Illness may be included in the process of dying as a factor that causes a weakening of life forces. In particular, the final stage of an incurable disease or old age assumes a dramatic course. Medicine tries to preserve the human dignity of the ill, disabled patients abased by pain. Art often returns to that moment of one's parting with his or her life. It is always doubtful whether the patient should be informed about the hopeless state.

"Euthanasia". Euthanasia is one of the problems of medicine that evokes much controversy. On the one hand, it is simply the case of putting an end to the patient's inevitable suffering. On the other hand, it seems to disregard basic moral laws. What inspires many artists and moralists is this gap between life – even a life full of suffering and pain – and death as an ultimate fact of eschatological, religious-philosophical meaning.

"Artistic approach". The artist's approach to the subjects connected with medicine is an artistic issue, but, at the same time, it is a question of the responsibility of truth. We usually have to deal with a raw, reliable realism which tries to faithfully represent all that is connected with illness and its treatment. Yet, for the sake of unrestricted imagination (apparent realism, mythical realism), we also find cases of departing from truth and reality. From formal poetry's point of view, it does not matter how much truth or how much fantasy there is. If, however, we take into consideration the stratum of general ideas of the work, the unrestricted fantasy diminishes the aesthetic value and makes the aesthetic experience shallow.

"Attitude". Attitudes towards the diseases have undergone a significant change in the second half of the 20th century. While in the past diseases were considered something shameful, private, and purely personal, now the conditions of illness are revealed (this concerns mostly persons holding political or social positions). Art also proceeds in this direction: we can find detailed descriptions of dying and of death. Drastic surgical operations are presented, films show suffering bodies and faces at the moment of death.

"Aesthetics of medicine". If this were expanded, it would aim at indicating the significant relations between art, the beauty of actual reality, and the broadly understood concept of medicine (i.e. medicine on the levels of diagnosis and therapy, in its psychological and sociological aspects). A particular discipline of this kind, however, would have to reach for the broader problems of philosophical anthropology and ethics. It would have to take into consideration the cognitive aspect of art ("cognitive aspects" meaning the elements of medical knowledge in a work of art). It would also have to take into consideration art as the element of self-cognition, the element which improves the physician's self-awareness and stimulates the therapist's active attitudes, to prevent the doctors from falling into the stagnation of routine, or from getting bored with the banal everyday hospital life.

In my academic contacts, while dealing with philosophical problems and the domain of values, I met physicians with a humanistic approach. It was they who indicated the questions about the dignity of doctors and patients, about the sense of their profession, and about the way in which a human being should refer to the tragic phenomenon of illness and suffering, about the way in which a feeling is born – the feeling that it is the human being's irrefutable obligation to respond to other people's needs in accordance with Hippocrates' moral creed.

Kraków

NOTES

1 J. Liebert, Wiersz o zagładzie, in: *Pisma zebrane* (Collected works), Vol. "Kołysanka jodłowa", Warszawa, 1976, p. 190.

2 The poem "Kołysanka jodłowa", ibid., p. 186.

GRAŻYNA PODRAZA-UCIŃSKA

MEDICINE AND ART

The customary association of the ideas of art and medicine, characteristic of European civilization since the times of ancient Greece, calls for a closer contemplation of the reasons for this situation. Questions about the relationship of art to medicine always recur. Is medicine an art? What is the art of medicine? What is the share of art in the medical practice? Is it creative? Can it be classified as *poiesis*?

There is also, on the other end of the spectrum, another group of questions. Does art perform a therapeutic function, and if so, to what degree? Is "*therapy using art*" a valid method of treatment, and what does it really involve? The never-ending need to answer these questions again and again seems to indicate that the reasons for pondering these issues are deeper than it often seems. I do not think that these reasons can be reduced to *self-gratification* as discussed by Władysław Szumowski[1] – with regard to the physicians who opt for intuitive cognition in medical practice – who forgot that a similar charge could be made against himself when he tried to classify all medical actions as scientific. Neither do I think that art must look for its *raison d'être* in its therapeutic role, nor improve on its prestige in that manner.

It would be more interesting and more productive – from the learning point of view – I believe, to explain why the interest in this issue has never died, in spite of the changes taking place in medicine and in art, resulting each and every time in a need to redefine their fundamental concepts and create new theories for both the medical practice and the art, and, consequently, to redefine the relationships between the two. I believe that a particularly significant question is whether and which similarities between these two types of human *praxis* are critical in never allowing these issues to become outdated, and whether the contemplation thereof is purely theoretical in nature, or also practical, and if so, whether therapy is one of its practical functions.

In order to answer these questions, let us first try to systematize the reasons why it is believed that the medical practice is, or should be at least to a degree, an art, and what aspects of the very notion of art could be involved here.

First, there is the treatment of the medical arts as a skill based on rational rules; ability, mastery, or artfulness of delivery is emphasized. This is, then, a rather general understanding of the notion of art, corresponding to the Greek

A.-T. Tymieniecka and Z. Zalewski (eds.), Analecta Husserliana LXIV, 103–111.

concept of *techne*, or the Roman one of *ars*, consisting of what we classify today as sciences, crafts, and the fine arts. As this original meaning of art was becoming obsolete and/or archaic, and the need for relating to a new set of beliefs was arising, a tendency emerged to start separating this set of rules from the practical delivery thereof, often construed narrowly in terms of technical skill limited to the examination of the patient and his/her treatment.[2] Consequently, this "set of rules" has been categorized as belonging to the sciences and the delivery – depending on the skill or mastery – of crafts or the arts. This value-based approach echoes the notions of *pure art* and *applied art*, so readily discussed around the turn of the century. It is the latter concept, combining practical sense and functionality with achievement – through mastery and perfectionism – of certain aesthetic values, primarily of what was referred to as functional beauty, which lies at the base of the differentiation between routine medical practice and that which is above average in medicine, therefore warranting the label *medical art.*

This idea of the medical arts was satisfactory primarily for the physicians whose beliefs were related to empiricism and positivism in which medicine was reduced to the natural sciences and engineering. However, the popularity of this concept, expectable in view of the development of medicine made possible through the achievements of general science and clinical research and through the inclusion into the medical practice of ever newer and better techniques, has eventually given rise to some disturbing situations. The most commonly mentioned are ineffectiveness and iatrogenic consequences of the treatment model based on reductional and patronizing precepts. A realization, clearer than ever before, of the fact that it is impossible to directly translate the principles of science to the standards of medical practice, let alone to that practice itself, is a consequence of these fears.

The above is impossible because the actions of a doctor are always directed towards solving an individual problem, while the principles of science are expressed in general terms. Additionally, the actions of a doctor are always performed according to a value system, and never for the sake of learning. A treatment also always involves psychophysical subjects (the doctor and his/her patient), and the relationship between the two may neither be reduced to that of a broken biological apparatus and its repairman, nor described in the language of the natural sciences. The *understanding* of the nature of this relationship requires the use of hermeneutic methods in their widest meaning and is in itself an important element of the treatment process that, in order to be effective, must always apply to the human as a whole, rather than focusing on his/her single fragment.

All this mandates that the notion of medical art be redefined in wider terms, in a way that will take advantage of the achievements of both science and technology, yet not lose sight of medicine's primary objective, i.e., the treatment and care of the sick and suffering. Incidentally, it seems that physicians should remember more often that it is the effectiveness of their actions, and not their scientific qualities, in which their patients are primarily interested. Scientific knowledge should not be opposed to the medical arts; on the contrary, it is its *sine qua non* condition. However, it may not be treated as a value in and of itself. Scientific knowledge is a significant element of the medical arts not because it provides a paradigm of science, but because it may be used for improving the effectiveness of treatment-related decisions, limiting the field of uncertainty and risk, and because the patient's belief in the effectiveness of the actions undertaken, and consequently his/her motivation to work with the physician, may be improved, his/her hope returned, and fears allayed through the scientific explanation of the pertinence of these actions.

This would then result in such a notion of medical art in which scientific knowledge is one of the tools used by the physician, used, demonstrably, towards various purposes. The question arises immediately as to whether other types of knowledge are also used, and if so, how that knowledge would have been gained. Two possibilities are traditionally indicated: the very controversial issue of intuitive cognition, and what is referred to as *medical experience*.

There is no room to delve here into the entirety of the philosophical issue of intuitive cognition. It should not, however, be entirely disregarded, as is commonly done by physicians involved in the discussion of this subject. It must be kept in mind that intuitive cognition is always direct, regardless of the type of intuition involved; therefore, of its very nature it may not be a type of reasoning (at most, it may supply reasoning with concepts). Doubtlessly right are those who find no place for intuition in what is referred to as *brief diagnostic deduction*, in which a physician, having observed a single, characteristic symptom, deduces what ails the patient.[3] This is rather what should be called the doctor's medical experience.

Does this mean that there is no room for intuition in medical practice? I do not think so. In order to answer this question, we should first decide what kinds of intuition we are talking about and what the subject of such cognition should be, and then determine whether such cognition is actually present in medical practice (any, or just the above average) and if so, what its role is in it. As far as I know, such comprehensive research considering various notions of intuition in philosophy, has never been done; furthermore, discussions have focused only on a selected facet of this notion, often forgetting even to define precisely that single facet; consequently, as could be expected, much confusion and many misunderstandings have arisen around this subject. The well-known discussion of the role of intuition in medicine that took place among the German doctors[4] in the first quarter of this century is riddled with these. Yet it would seem that the popularity of Bergson's concept of intuition as cognitive power, capable – through a kind of emphatic bond – of capturing in a synthetic and dynamic manner, the uniqueness and the completeness of the qualitative accouterments of individual being, is the very concept that could try to explain the aspects of medical art applicable to the patient's unique individuality. Because if a physician is to effectively influence the patient's case, the principal prerequisite for that is to know it first, which is made difficult by both the shortness of time available for that and the fact that the sick person's individuality is hidden under the mask of a patient. The ability to penetrate it and undertake an actual dialogue, or if this proves impossible, to apply other forms of non-verbal communication (also those developed in the arts) is one of the most significant elements of medical art.

Another type of intuition that should be addressed in the context of medical art is phenomenological intuition, and particularly its Scheleresque version, which – through the intentional nature of *emotional functions* and *emotional acts* – allows for the cognition of values: not only of their existence and the degree of their value-based nature, but also of their quality, types, and hierarchy. If a physician is to direct his/her actions by values and achieve (or reinstate) values, the issue of the method of the cognition of values must be undertaken. It is not enough for a doctor to observe the learned ethical standards, because these very standards, and particularly the most important one of defining the duty to act for the patient's benefit, need further interpretation. This interpretation depends on the breadth of the range of values that may be sensed by the physician, not only the vital and ethical ones but also other types of values, and on the physician's perception of the hierarchy thereof. It is impossible in medical procedures to carry out and protect – except by accident – values that one cannot recognize. It is also impossible to correctly interpret the principles of the patient's good without understanding which values are particularly significant and should be the first to be protected. The allowance for the above is often very difficult, as it requires the physician to act in a manner that is non-routine, creative, and that sometimes carries even an increased risk of failure; but these are the actions, carried out because of the values protecting the patient's autonomy and focused on not only his/her life, but also its quality, that deserve being termed *medical art.*

The third type of intuition, associated with what is referred to as *medical thinking*, is, unlike to the former two, intellectual in nature. This would be the ability, or skill, manifested particularly under the conditions of a high degree of uncertainty, of choosing the right diagnosis and treatment methods. This refers to the uncertainty arising of course not from the physician's lack of education, but from the medical issues' high degree of complexity and the awareness of the limited nature of the very scientific knowledge itself, or from the unavailability of some critical information about the patient. In this case, the intuition would consist of the ability to create a synthetic picture of the situation, directed at capturing the significant relationships, based on the use of non-specific information. In conjunction with the imagination construed as a field of research for arising possibilities and consequences of both diagnostic judgements and treatment techniques, intuition would be an important element of the decision-making process. Intuition and imagination are here construed as both the necessary elements of creative thinking and medical action that are critical to their effectiveness under particularly difficult and atypical conditions, and the basis for separating the creative element of a doctor's work from the routine. This is the approach proposed by Jegerman.[5]

Let us now take up the notion of *medical experience*, since it seems that the manner in which this extremely complex issue has been understood reflects well the change that has taken place in the notion of medical art from the times of the positivist paradigm's unchallenged rule to its contemporary concept.

Medical experience, as the sum total of the empirical knowledge gathered through practice and the technical skills mastered to perfection, was valued by physicians with positivist attitudes for two main reasons. First, because the collected empirical knowledge was supposed to make scientific propositions more probable and provide the material for calculations of statistical probability. It was to supply the doctor him/herself with the justification for diagnostic and treatment related decisions through *subjective probability*. Secondly and consequently, this would impact the speed of decision-making and improve the probability that the right decisions were being made. The physician's technical skills assured the speed and effectiveness of the procedures performed. However, both these extremely important issues, particularly with regard to doctors who are overworked and whose time with any single patient is largely limited, emphasize only the non-creative moments in the medical practice and those that correspond to the notion of medical art as a craft performed with skills mastered to perfection.

I believe that now, when the ideas of creativity and novelty are so fashionable and so often – unfortunately – overused, one should look at medical experience from this point of view. It seems that in medical arts, not unlike in the fine arts, the achievement of a high degree of proficiency is, for some, an ultimate objective after which there is nothing but *duplication*, while for others the realization of the limits of possible action delineated each and every time by the level of the techniques used becomes an inspiration to find new solutions and new approaches. If one were to use a parallel to painting, Canaletto could be an example of the former type of attitude, while Titian would be of the latter. There is, of course, room for both these types of attitudes both in art and in medicine, but we tend to value the second type, open to new search, more highly, precisely because of its truly creative nature. So, medical experience would be valuable here as a condition of the development of the medical art. However, to ensure that this is a true development, and not a mere pursuit of novelty at any price (a condition seemingly plaguing both modern medicine and art), it seems necessary to call once again upon medical experience, this time construed as a developed ability for medical thinking, in order to ensure the correct, critical evaluation of the benefits and risks and to make a choice based on the values to be realized by the new method.

This situation may often lead to a value conflict, such as the need to sacrifice the innovator's fame, or, on the other hand, to take the chance of being charged with undertaking undue risks. The development of the correct hierarchy of values and the courage to carry out higher values also belong within this wider meaning of the notion of medical experience corresponding to the contemporary idea of medical art emphasizing its creative elements. This very meaning of the concept of medical art justifies the question of whether it is not more of a *poiesis* than a *techne*, even though the issue of the resultant healing is still somewhat dubious.

Let us now return to the questions posed at the beginning of this paper and let us try, conscious of the multiple meanings and the complexity of defining the ideas discussed herein, to indicate, even if only tentatively, the range of possible answers. The questions pertained to the possible similarities between medical art and art construed as artistic creation, similarities that would justify the use of the word *art* with regard to both, and at the same time indicate the reasons for including the arts in the therapeutic process.

I shall focus on the more general of these. Among the most significant I would first see the common sources of both of these types of human activity: the realization of the deficiencies and imperfections, and the realization of the fact that – from the perspective of value – *something that should exist does not* or that does exist is not *how it should be.* To this realization corresponds, in the emotional sphere, our suffering and the suffering of others, and the desire to relieve or at least lessen it. The suffering is here construed in general terms, not only as the sensation of pain, but also in terms of psychic or spiritual suffering. Medical art doubtlessly grows out of the suffering and the desire to relieve it; it is however impossible to prove at this moment that the situation is similar with regard to fine arts (I have addressed this issue in more depth in an article, *Art and Suffering*).[6] But in order not to state what I cannot corroborate, I could give the example of the compensating function performed by art or its ability to produce *catharsis*, both observed by people with medical training (Adler, Aristotle), and both used in therapy.

I see the second similarity in the fact that the range of the values considered and realized *in concreto* is exceptionally large in both these fields. Virtually all types of values, vital, hedonistic, ethical, aesthetic, and metaphysical, are included at least to a degree, even though they appear in varying arrangements, since those which are important in the medical arts will differ from those critical to the fine arts; however, the more these can be protected by the physician, returned, and brought about, the more his/her work deserves to be called medical art;[7] similarly, the more these are realized by an artist in a work of art, the closer he/she is to creating a masterpiece.

The third similarity – actually, the preceding one seems to be its outcome – lies in the fact that the work of both the doctor and the artist engages his/her entire personality in the physical, psychological, and spiritual dimensions, and its aim is to act on the entire personality of the patient or art viewer. This gives rise to a number of similarities related to the structural features of both the *therapeutic* and *aesthetic situations*, as well as the means used by the physician or the artist for the creation thereof. The wealth of forms of verbal and non-verbal communication developed in the arts, the continuous development and improvement of various forms of expression and symbolization, and the various ways of molding the image of reality may all provide a valuable source of inspiration for the development of medical art. This justifies the proposition to include art, to a larger degree than heretofore, in the process of patient treatment (not only psychiatric), and to treat intensified contacts with art as an essential element of occupational development in the medical profession.[8] However, in order to achieve the latter, the viewer's relationship with art must not look to the expectation of an elementary aesthetic pleasure, but must focus on a knowledge-based type of conscious perception directed toward giving justice to the work of art, grasping the entire wealth of values presented in it, and toward the ways and methods of expressing them, its formal construction, and the cognitive content it delivers, perceived correctly. This type of relationship with art may not only arouse the doctor's sensitivity dulled by constant contact with the negative aspects of human existence, but may also widen its confines, assist him/her in developing his/her imagination, deliver from schematic thinking, and broaden his/her means of expression and methods of symbolization.

And finally, the fourth and most obvious similarity pertains to the requirement of mastery of skill and artfulness common to the physician and the artist. I have deliberately chosen to set these two apart, since the mastery of skill does not in itself guarantee artfulness; the latter does not take place until the former is used in an individual manner, fine-tuned each time to the needs of a specific art piece or treatment course. This takes place, for example, when a physician decides on the technical methods of pharmacological procedure, adjusting them on an individual basis to the discrete properties of the patient's body, his/her psychic situation, and the limitations mandated by such factors as the nature of the patient's work.

I would like to conclude with a few words on the possible therapeutic function of the contemplation of medical arts. I believe that even though these thoughts arise primarily from the physicians' concerns about the objectives, sense and value of their profession, they also reflect wider anxieties and fascinations characteristic of their times. The elements of medical art that are particularly valued at a certain time are singled out, and the public is being assured of the need to respect the values generally perceived as particularly threatened. It was like that in the nineteen twenties and thirties, during the time of fascination with the seemingly unlimited possibilities of science and engineering; it is like that now, in the time of a crisis of trust in science and a fuller awareness of the risks involved in the uncontrolled development of technology. Therefore, it seems that one of the functions of a reflection on medical art is to allay patients' fears and reinstate their faith and trust in medicine, thereby creating a situation conducive to successful treatment.

One should only hope that the patients' belief that their most prized values shall truly be respected will be reflected not only in theory, but likewise in practice; today these values are primarily: respect for the patient, for his/her autonomy and dignity, and for a high *quality of life*.

Jagellonian University

NOTES

[1] Szumowski, W., 1927, "Medycyna jako nauka i jako sztuka" (Medicine as a science and as an art), *Archivum Historii i Filozofii Medycyny*, vol. 7, n. 2, p. 200.
[2] Ibid., p. 204.
[3] Ibid., p. 203.
[4] Discussions concerned the following works: Schweningen, E., *Der Arzt*, 1907; Hönigmann, G., *Das Wesen der Heilkunde*, Leipzig: 1924, and *Das Problem der Ärztlichen Kunst*, Giessen, 1922. In Polish: Gawliński, J., "Poglądy Hönigmanna na medycynę, jej istotę, rozwój i podstawy filozoficzne" (Hönigmann's view of medicine, its essence, development, and philosophical foundations), *Archiwum...*, 1926, vol. 5, pp. 278–282.
[5] Jegerman, K., "Intelektualna biegłość lekarza" (The physician's intellectual proficiency), *Przegląd Lekarski*, 1984, vol. 41, no. 10, pp. 595–600.
6 Podraza-Ucińska, G., "Suffering and Art", *Suffering as Human Experience. Proceedings from the VIth Jagiellonian Symposium on Ethics*, Kraków: 1994.
[7] Dąbrowski, S., "Rozwazania o sztuce lekarskiej" (Considerations regarding medical art), *Archivum...*, 1989, vol. 52, pp. 35–46.
[8] Święcicki, H., "O estetyce w medycynie" (On aesthetics in medicine), *Księga pamiątkowa XI Zjazdu Lekarzy i Przyrodników Polskich* (Proceedings from the XIth Assembly of Polish Physicians and Natural Scientists), Kraków, 1911; Gryglewski, R., *Medicine: How Much Science, How Much Art?* Kraków: 1995, Jagiellonian Medical Research Center.

PART II

CONCEPTIONS OF THE HUMAN BEING

HENK TEN HAVE

THE ZAPPING ANIMAL: OSCILLATING IMAGES OF THE HUMAN PERSON IN MODERN MEDICINE

1. INTRODUCTION

The life of modern man was frequently likened to the pilgrimage-through-time. The itinerary of pilgrims is drawn in advance by the destination they want to reach ... and everything they do is calculated to bring them closer to the goal. The pilgrim is consistent in choosing every successive step, conscious that each step matters and the sequence cannot be reversed. Today's men and women can hardly treat their life as a pilgrimage, even if they wish to.... The life of the men and women of our [postmodern] times is more like that of the tourists-through-time: they cannot and would not decide in advance what places they would visit and what the sequence of stations would be; what they know for sure is that they will keep on the move, never sure whether the place they reach is their final destination [Bauman, 1994].

The human condition of the present day is characterized above with a comparison between modernity and postmodernity in the words of Zygmunt Bauman, a scholar of Polish origin and professor emeritus of sociology at the University of Leeds. In his recent book, *Life in Fragments: Essays in Postmodern Morality* [Bauman, 1995], Bauman analyses four types of intertwining and interpenetrating images of postmodern man: the human person as the stroller, the vagabond, the tourist, and the player. These images not only characterize various lifestyles, but also lead to different self-interpretations of human beings, with far-reaching implications for human relationships and moral responsibility. Within a world that has replaced durable objects with disposable products, and guided by the determination to live one day at a time as the ideal of rational conduct, not committed to controlling the future but refusing to mortgage it, postmodern life strategies are not focused on identity-building, but on the avoidance of being fixed. When the horror of being bound and fixed is the most important determinant of the self-understanding of contemporary human beings, the scientific agenda of philosophical anthropology should focus on analyzing the floating and discontinuous lifestyles, rather than identifying and delineating more or less cohesive and continuous ideas or concepts of man. Postmodern life is incoherent; every identified life strategy conveys only part of the story, almost never integrating into a totality.

A.-T. Tymieniecka and Z. Zalewski (eds.), Analecta Husserliana LXIV, 115–123.

This common diagnosis of the postmodern predicament, described and clarified by Bauman, provokes questions such as: What does this diagnose? How adequate or correct is it? What is the significance of an anthropology of fragmentary and discontinuous human life for a philosophy of medicine? Furthermore, there also is the preliminary question of how such a diagnosis is possible. If fragmentation and discontinuity are the essential hallmarks of human life, from what perspective are we able to perceive and recognize the postmodern image of the human being, we ourselves, as fragmentary and discontinuous?

2. THE SIGNIFICANCE OF PHILOSOPHICAL ANTHROPOLOGY

In the final paragraph of his historical overview, Szumowski has proposed a definition of the philosophy of medicine, and distinguished several subsections of the discipline: logic and ethics, and perhaps epistemology, psychology, and medical metaphysics [Szumowski, 1949]. It is curious that he did not explicitly refer to philosophical anthropology. On the one hand, it is obvious that the problem of man has been a continuous object of reflection during the long history of philosophy of medicine. In many varieties, and from different perspectives, medicine, and therefore also the philosophical self-understanding of medicine, have attempted to answer the basic question: What is man? Time and again, medicine, as Szumowski shows, has grappled with issues of organicism and mechanism, holism and reductionism, vitalism and materialism. Some understanding of man is crucial for the mission of medicine. Whether it is the view of the person as a whole or the view that man is a sophisticated collection of particular organs, some view of the human being is essential to demarcate and define the object as well as the methods of medicine as science and practice. Man himself plays a central role in medicine, as one subject to and suffering from disease; the human being is the object of interest as well as the subject of study. As, for example, William Inlow [1946] has indicated, when these truisms refer to essential characteristics of medicine, a philosophy of medicine cannot be developed without philosophical anthropology. In a very basic respect, anthropology necessarily is a presupposition of a philosophy of medicine. Anthropological questions have been pervasive throughout the history of this discipline, perhaps so pervasive that they did not need to be identified separately. A philosophy of medicine in this perspective can be regarded as an uninterrupted essay on man. One of the consequences of this view has been presented by Szumowski in a 1937 essay. In medical education, students need a coherent framework of knowledge, practice and values; they need to envisage medicine as a unity. But then, medicine and its representatives need to proceed from a unified image of man; they cannot continue to divide the patient into a disease and a person, into a body and a psyche, into an entity as a whole and its constituent organic parts [Szumowski, 1937].

But, on the other hand, medicine itself can also be considered as man's effort to understand himself. Medical science not only generates knowledge and facts about human nature, but also contributes to particular concepts of man. Medicine, perhaps more than the other sciences, helps to produce and construct such concepts. Ernst Cassirer has pointed out that in science man can do no more than build up his own universe. If this is true, then medicine adds something more: through contemporary medicine man can build himself up as part of that universe. Medical activities articulate what concepts and ideas of man contemporary human beings attempt to realize. Medicine, then, can be one of the preeminent examples for learning more about the philosophical ideas about the human person that prevail in present-day culture. To medicine, Cassirer's statement can be applied, that... "Man's outstanding characteristic, his distinguishing mark... is his work. It is this work, it is the system of human activities, which defines and determines the circle of 'humanity'" [Cassirer, 1976, p. 68]. From this perspective of medicine as self-expression of the human being, it is urgent to reflect upon the images of the human exemplified, constructed and realized through medical activities. Philosophical anthropology no longer can remain an implicit presupposition, but needs to be developed into an explicit, articulated and analytic activity.

3. THE ANTHROPOLOGICAL PHASE OF A PHILOSOPHY OF MEDICINE

Particularly since 1870, there has been a rapid growth of medico-philosophical literature. It is argued that in the thematic development of a philosophy of medicine since that renewed starting-point, three traditions can be distinguished: an epistemological, an anthropological, and an ethical tradition [Ten Have, 1995]. This distinction can help to clarify the continuity of the basic themes, preoccupations, and motivations within the discipline of philosophy of medicine. The few historical studies available, notably Szumowski's [1949], do not demonstrate any evolution of the discipline; one may easily have the impression that there has been a succession of philosophizing doctors and a rather accidental accumulation of books and articles. Retrospectively, it might even appear that philosophy of medicine is a very recent affair and that the present preoccupation with medical ethics shows a marked discontinuity with earlier efforts to philosophize about medicine. The present-day domination of medical ethics is, however, continuous with the various traditions of philosophy of medicine in that it is moved by the same commitments and fundamental problems. This continuity is manifest, especially in relation to the tradition of anthropologically oriented medicine which flourished prior to the current interest in ethics, particularly in Germany and the Netherlands.

From the beginning of this century, physicians and philosophers of medicine such as Buytendijk, von Weizsäcker, von Gebsattel, Plugge and Christian sought to redefine and reinterpret medicine as a science of man, applying and developing ideas from several contemporary philosophical schools, particularly phenomenology, existentialism, and philosophical anthropology [Ten Have, 1990].

The basic ideas of this movement can be summarized under three headings: (1) the rejection of anthropological dualism, (2) the notion of medicine as a science of the human person, and (3) the necessity of a comprehensive understanding of disease.

The first set of ideas has developed from a critique of the dualistic ontology usually underlying medical theory and practice, subdividing human beings into a physical and a mental compartment. Medicine, of course, has profited enormously from this subdivision, but it has also restricted itself to the human body by studying and explaining the body's physico-chemical machinery. This approach has a tendency to reduce the human being to a specific animal species, and the human characteristics of the human body to its physical level of being. Anthropologically oriented physicians argue that any demarcation between body and mind is artificial. Dualistic thinking, moreover, not only involves a reduced image of the human person, but it also has a more general tendency of making a stringent distinction between object and subject. Such thinking erroneously suggests that there is an objective, real world, independent from an isolated, individual subject. It also leads to an almost exclusive preference for the methods of the natural sciences in the context of health care. These methods are focused on intervention, control and manipulation, introducing the technical point of view of the engineer into the domain of disease and suffering.

The second set of ideas aims at developing a more appropriate understanding of medicine as a science of the human person. The point is not that the causal thinking and technical approach of the natural sciences should not be allowed in medicine; on the contrary, they are valuable, but they are not sufficient. Medical thinking and practicing should not restrict itself to these scientific methods, because they cannot grasp what is essential to human beings. As a living organism, every person constitutes a whole, a meaningful entity, which is disconnected and disintegrated in abstract, analytical approaches. What we need, for example, is an anthropological physiology, not only explaining those aspects of living organisms that can be analyzed from a mechanical point of view, but also clarifying the intrinsic meaningful connections and interrelations of bodily phenomena as specifically human, as well as the significance of human experience and conduct [Dekkers, 1995]. If medicine wants to evolve into a science of the human person, it should overcome the distinction between the objective and the subjective, introducing the subject into medicine itself. Buytendijk, one of the representatives of this anthropological tradition, summarizes his position as follows: If medicine is not objective, it is impossible; but if medicine is only an objective science, it is inhuman [Buytendijk, 1959].

What is true for the methodology of medicine also applies to the medical understanding of disease. Anthropologically oriented doctors developed their third set of ideas from a criticism of medicine's approach to patients. Focusing on the causal mechanism of disease, medicine can understand the ill person, because explaining disease also refers to the problem of the significance of a symptom, the meaning of a particular complaint [Welie, 1995]. Science-based medicine in fact hinders the insight that disease has meaning [Welie, 1994]. For patients, relevant questions do not so much refer to the pathogenesis and pathophysiology of the disease, as to the anthropological query: "Why am I suffering here and now?" Being ill is primarily an existential category; only secondarily, can we make any differentiation between organic phenomena, subjective complaints and personal suffering. Being ill is a way of being a human person. From an anthropological perspective, human beings do not only have their lives, but also give expression to them; they do not merely have their bodies, but also are their bodies, at the same time. From that same perspective, it is not only the case that persons have their disease, but that they also make their disease as a response to their own individual existence. In this perspective, disease is not a negative event, a blind fate, waiting to be eliminated from the world by technological intervention; the important thing is what we make of it, whether we consider it as an occasion to reconsider and improve our life.

4. MAN AS PILGRIM

In their writings, representatives of the anthropological movement in medicine tried to reflect upon human existence in its concrete specificity and ambiguity. Instead of starting from or working toward an ideal image of the human being, they attempted to identify what is anthropologically characteristic and common to all human beings. But at the same time, they were very much aware that any image was too abstract and "clean", because in everyday reality the specific individual was always changing, pluriform, and was not fully described by the designed image of a person. Given this theoretical point of view, advocates of anthropological medicine have not presented a clear-cut theory that has been defended and elaborated upon by all representatives.

What is important, however, is that they have helped to articulate, and criticize, the image of the human person that underlies, justifies and stimulates much of everyday medicine: a universalistic and reductionistic image. In this image, human beings can be understood by analyzing and studying anatomical structures, physiological functions, pathological aberrations, biochemical complexities or genetical locations and dislocations. Such criticisms are not at all outdated or anachronistic; reductionist images tend to reappear with every new and promising scientific development.

The problem is that the antidote, in other words the alternative image presented in anthropological medicine, is now itself questionable. For the anthropological critics, human beings are essentially dynamic beings adapting to the world; they reconstruct and redefine their identity in order to keep it solid and stable; they reconsider their life as a project, an individual task to accomplish the goals that have been set, and to give meaning to the various stages and experiences. This is the image of human life as pilgrimage, described by Bauman [1994, 1995]. For man as pilgrim the true place is always some time, some distance away. What makes his life worthwhile is the distance between the true world and this world; he is destined to be elsewhere. Because life has been transformed into a pilgrimage, it receives meaning. It is the destination that makes a whole out of the fragmentation, that lends continuity to the episodic. Human life therefore is a continuous story, and although it is an individual project, it is carried out in an orderly, determined, predictable world. The image of the human being as pilgrim, which has more or less implicitly motivated anthropological physicians for a long time, has specific characteristics (such as goal-directness, the deliberate choosing of successive steps in life, a slowly changing world requiring man to adapt, and unity and continuity of the life project) which no longer apply, according to Bauman, to the postmodern situation.

5. MAN AS TOURIST

The postmodern human being is characterized by "endemic non-finality" and "non-fixity" [Bauman, 1995: 78 ff]. There is no final state, no state of perfection which may be realized in search of improvement; life has only local and transitory achievements. The defining feature of postmodern existence is the fear of being fixed. Persons avoid fixation, commitment and stable relationships. They like to keep their options open. For them, the ultimate freedom is celebrated in the activity of zapping. The postmodern individual has ultimate control over the images he or she wants to produce, and can change them at will. Contingency, episodicity and fragmentation are the marks of human life. In a very basic sense, the image of the human being is merely and nothing more than an image; the image is also perpetually changing. Bauman therefore prefers to speak of lifestyles and life strategies, rather than of images of man.

This world of lifestyles no longer seems hospitable to pilgrims. The anthropological critique of modern medicine is not feasible anymore, in this world of dissipative structures that has lost its definiteness and continuity, this human reality that exists as a series of episodes or events without a past and with no consequences. The postmodern person is best characterized by the image of the tourist. This person, in the words of Bauman, is "a conscious and systematic seeker of experience, of a new and different experience, of the experience of difference and novelty" [Bauman, 1995: 96]. For man as tourist the world is domesticated; it is obedient to his wishes, it is made to please. The most important thing is that the world should be structured by aesthetic criteria. For man as tourist, it is not clear in the end, where home is; having a home becomes a mere postulate. The tourist belongs nowhere, but dreams of belonging.

We do not need to agree with this diagnosis of postmodernity in order to recognize the consequences. The postmodern person is an interest-seeking subject. For him or her, the objects in the world are not relevant as entities in their own right. What matters is whether they are pleasing or not pleasing, satisfactory or unsatisfactory. The tourist does not want to change the world, he wants to lead an enjoyable life. Reality cannot challenge him; it does not provoke him towards rectification, improvement, transformation. Postmodern life strategies are furthermore focused on rendering human relations fragmentary and discontinuous. Doing so, they promote the distance between the individual and the other; the other person moreover is primarily considered as the object of aesthetic, not moral, evaluation. The effect of these postmodern strategies of disengagement and commitment-avoidance is the suppression of the moral impulse. What transforms experiences into moral experiences has much to do with responsibility for the other, engagement in the fate of the other, and commitment to the other person's welfare.

What postmodern literature makes clear is what essentially is the relationship between anthropology and ethics. Instead of the primacy of anthropology, with ethics based on anthropological theory and concepts of the human being, a philosophy of medicine should focus upon explicating the morality underlying anthropology itself.

The question is not whether the image of the human being as a pilgrim or a tourist is an adequate description of the present-day situation. To verify or falsify such a description is a matter of sophisticated sociological studies. Even with the most complicated study designs, however, it seems almost impossible to make the description into a general characterization of present-day human beings, since its truth value is so dependent upon the cultural and social circumstances; what is a true sociological representation of human beings in Texas most probably is not true for "postmodern" human beings in Catalonia or Calabria.

The more interesting question, however, is a normative one: What underlying notion of morality makes it possible to diagnose the human situation using such images? The possibility–condition seems to be a basic understanding of morality, not as something we choose, but as a fundamental predicament we are already involved in before we even start to reflect upon it. Morality is choosing us, because we are primarily social beings. Ethical views are articulated and explicated because we are in a moral relationship with other human beings appealing to us. The face of the other makes us moral beings whether we like it or not, whether we choose to act accordingly or not. This interpersonal character of morality is the implicit philosophical background of the postmodern diagnosis of fragmented and discontinuous lifestyles. The diagnostic result is that moral views are considered objects of individual choice, items on the marketplace that we prefer or not, and exchange whenever we like. The outcome of the diagnosis, however, should not be confused with the possibility or condition for making the diagnosis. Morality is a social affair. Its interpersonal character makes it possible to scrutinize and criticize individual moral choices. Tourism as an individual experience is only possible within a huge network of human cooperation. It annihilates itself when individual preferences destroy the quality of the network.

Catholic University of Nijmegen

BIBLIOGRAPHY

Bauman, Z. *Alone Again. Ethics after Certainty*. London: Demos, 1994.

Bauman, Z. *Life in Fragments. Essays in Postmodern Morality*. Oxford: Blackwell, 1995.

Buytendijk, F. J. J. "De relatie arts-patient". *Nederlands Tijdschrift voor Geneeskunde* **103**, 1959: 2504–8.

Cassirer, E. *An Essay on Man*. New Haven and London: Yale University Press, 1976 (original 1944).

Dekkers, W. J. M. "F. J. J. Buytendijk's Concept of an Anthropological Physiology". *Theoretical Medicine* **16**, 1995: 15–39.

Inlow, W. D. "Medicine: Its Nature and Definition". *Bulletin of the History of Medicine* **19**, 1956: 219–273.

Szumowski, W. *L'Histoire de la médecine et la réforme des études médicales*. Cracovie: Librairie Gebethner & Wolff, 1937.

Szumowski, W. "La Philosophie de la médecine, son histoire, son essence, sa dénomination et sa définition". *Archives Internationales d'Histoire des Sciences* **2** (9), 1949: 1097–1139.

Ten Have, H. A. M. J. "The Anthropological Tradition in the Philosophy of Medicine". *Theoretical Medicine* **16**, 1995: 3–14.

Ten Have, H. A. M. J., Kimsma, G. K., Spicker, S. F. (eds.). *The Growth of Medical Knowledge*. Dordrecht: Kluwer Academic Publishers, 1990.

Welie, J. V. M. *In the Face of Suffering. Prolegomena to a Philosophical Foundation of Clinical Ethics*. Nijmegen: Universiteitsdrukkerij, 1994.

Welie, J. V. M. "Viktor Emil von Gebsattel on the Doctor-Patient Relationship". *Theoretical Medicine* **16**, 1995: 41–72.

CHRISTIAN BYK

NEW BIOMEDICAL DEVELOPMENTS AND THEIR IMPACT ON THE LEGAL THEORY AND PRACTICE OF THE MIND-BODY RELATIONSHIP

As the Holy Scriptures state: “Spiritus promptus est, caro autem infirma.” Can the difference of nature between our Mind and our Body lead to the disaggregation of our Being?[1]

We shall not attempt to answer this question, which has recurred throughout history, but we are sure that an answer could help us today to re-examine the legal position of the human body. Indeed, we think that the time has come for this question to be raised again, because recent biomedical research has revealed that it is possible not only to treat, but also to modify and transform, the human body and its components.

In 1806, a professor of law taught: “Man is both body and mind and each of the two needs the other.”[2] As components of the human body, such as blood, gametes and organs are now “in circulation” outside the human body and can also be transformed and manufactured, we are about to lose our legal certainties about our notion of the human being (see I).

However, the emergence of a specific law for the human body will not necessarily lead to the rejection of the old idea of one integral human being (mind and body), if appropriate concepts and techniques are used to clarify the new status of the human body (see II).

I. ENDING WITH LEGAL CERTAINTIES

In modern history, the human body has never been regarded as a distinct legal concept. However, the body is the material which gives rise to recognized legal concepts such as the human person, and hence is unique, and hence enjoys a special legal position.

A. *Definition and Nature of the Human Body*

1. Definition

Prof. R. Dierkens, an eminent expert in medical law, considers that

> Each product of conception reaching a certain degree of development is a body. This body is neither a man nor a person because the body is not an abstract philosophical or legal concept. It is essentially an anatomical and biological reality.[3]

A.-T. Tymieniecka and Z. Zalewski (eds.), Analecta Husserliana LXIV, 125–133.

Therefore, if the biological aspect of our existence is a consequence of (possessing) a body, then it follows that the legal aspect of our existence is a consequence of (the existence of) the law.

2. The Legal Nature of the Human Body

J. Galloux states that "the division of our sensible world by the law into two main categories, persons and chattels, is a major evidence."[4] Unfortunately, the human body does not fall neatly into either of these categories, and for this reason, as Dean J. Carbonnier noted: "The human body never appears as such in the Civil Code."[5]

However, lawyers have always considered that the human body should be classified among persons. Once again Carbonnier states that "the human body is the substratum of the person,"[6] and Dean G. Cornu asserts that "the body is not a chattel; it is the person him/herself"[7] and as a necessary "substratum of the person, the body is what exists before the mind becomes stronger and what remains when the mind is diminshed."[8]

A person is understood in the Civil Code as a thinking non-physical entity. The philosophy of those who inspired and drafted the code was spiritualist in the sense that they considered our personality to be more a will than a body. But that primacy of the will – the mind – over the body does not exclude the presence of the body in the law. There is no marriage without the body, no procreation (not even artificial procreation) without bodies or gametes.

Therefore, the law considers that humanity exists only because there is a body which is the instrument of this activity. The body and the mind make a whole which testifies to our humanity. But, today, can we still consider this as the truth "while procreation and death are manipulated, while new infringements to the body integrity are accepted or even claimed by physicians?" Is the human body moving from the category of persons to the category of chattels? Is the human body losing its sacred character?[9]

B. *The Human Body Protected by Its Legal Qualification*

Because the human body was regarded as part of a person, it benefitted from the protection granted to the human person: protection against violation of integrity, protection against immoral arrangements.

1. The Rule "Noli Me Tangere"

The primacy of the rule "noli me tangere" is fundamental both in private relationships and in the relationship of each human person with society and the state.

a) In some countries, such as France, the rule is so absolute, that it is usually impossible for a civil court to force someone to submit to treatment, or to blood sampling in a paternity case. But would not the possibility now presented by new scientific techniques using biomolecular tests to certify, for example, paternity, make a considerable change in the legal approach to this question? In particular, can the right to know who one's parents are, which is protected under the UN Convention of the rights of children (the initiative for which came from Poland), be regarded as justification for a compulsory recourse to such techniques?[10]

In other areas such as organ transplantion, in many countries legislation has already been passed which accepts and regulates this practice. The legislation covers living as well as dead donors and is obviously a violation of the integrity of the donor for the benefit of the health or the life of the recipient.

b) Regarding the protection of the individual from State interference, we can observe that since the Enlightenment, the legal concept of personhood has been extended; and categories such as slaves and the dead under civil law have not disappeared, making those concerned benefit from the protection of the "noli me tangere" rule.

However, infringements to this rule have always been possible when there have been legitimate reasons such as public health, or the prevention of criminal offences. But in democratic countries, these possibilities are strictly limited and controlled by the judiciary.

Moved by strong scientific, medical and sometimes industrial incentives, biomedical research on human beings has now opened the way, through genetic engineering, to a possible reshaping of the human body, but also to a mastery of procreation and heredity which could affect the entire human species. The legislative trend in the United States is towards allowing more intervention in the human being. (The United States is at the forefront of the above techniques.) Counter-arguments in favour of protecting the human body now rely less on the "noli me tangere" principle and more on the principle of the autonomy of each individual and the idea of free informed consent. But should we not protect the individual even against his/her will?

2. The Limits of the Right of Autonomy

The fact that a person can plan how his/her body will be disposed of after death implies recognition of the fact that the will can be separated from the body on which it acts. But this reasoning should not allow a complete reification of the human body, because the body and the human person are a whole. For this reason, in many countries the law, in the name of public order, demands that the individual – I mean his/her body – should be, if necessary, protected against his/her own will. This is, for example, the meaning of the "Res extra commercium rule" which is incorporated in the civil code: the human body should not be submitted to contract law. Of course, exceptions or clarifications have usually been admitted in order not to "push too far the fear of 'sacrilege'" – Professor Carbonnier.[11]

On the one hand, the clarifications relate to the illegality of contracts concerning the human body. Contracts in which the body is moved by will, such as contracts of transportation, are not illegal. On the other hand, a few exceptions have usually been admitted when there is a social interest – paying a woman to give her milk to a child – or when the violation is very limited – selling hair, for example.[12]

The problem today is that the biomedical revolution has so greatly extended the scope of uses of components of the human body that we could *de facto* wonder about what remains of the principles I have just mentioned.[13] Elements of the body, highly valuable in terms of symbols such as blood, placenta, gametes and genes, have become matters of scientific and medical research; they can even be part of industrial applications.[14] Organs are also widely used: a kidney or a piece of liver can be collected from a living person to be successfully grafted on to a recipient. In some cases, parents have decided to procreate in order to create a compatible donor for a child waiting for a transplant!

In surrogacy, it is the body as a whole which is solicited for a specific function – reproduction – but in drug research, the entire body functions for the purpose of the experiment "necessarily immoral but morally necessary," as J. Bernard said.

Hidden by the law under the concept of a person, but revealed by our biomedical techniques, will the human body come into "legal existence" as such only to be utilised and dismembered for the benefit of the individual and the community, each claiming a right to it? Will then the law, and particularly the legislation recently adopted in the biomedical field, "dig a grave for" the body? I fear that the law will lose its mind in doing so, but as J. P. Kauffmann said: "My body, this is the mind of my life."[15] In that case, what kind of law should help to define and protect the human body?

II. QUESTIONS ABOUT THE NECESSITY AND FEASIBILITY OF ESTABLISHING A LAW OF THE BODY

We acknowledge that the bioethics debate is a good opportunity to remind us of, and sometimes to search for and validate, the founding values of our life and our society. In particular, can we find in our law today elements which will help to solve these two questions: Do individual rights conflict with respect for the human body and human dignity? Is public interest in using the human body a reason for less liberty and the wasting away of our human condition? Are we simply becoming cannibals, as J. Attali suggested?

As a lawyer, I would now like to focus on two risky alternatives: The first alternative is to reject the system founded on respect for individual rights. The second is to consider that our law does not have to be renewed in its concepts, notions, methodology and systematisation.

A. *Protecting and Extending the System of Individual Rights*

Are the critical arguments developed against the concept of individual rights strong enough to suggest that we should get rid of the whole system?

1. How Strong are the Critical Arguments?

Is it true that the new biomedical techniques – because they are challenging fundamental notions such as individual autonomy, human personhood, and species – are destabilizing our legal system and concepts?[16]

The fact is that with the biological revolution,

> individual rights are no longer those of a citizen, but those of a person as a being made of flesh and blood. Thus it is impossible not to raise the question: is there any legal idea of what a human being is at the end of the 20th century, I mean a man in his globality?[17]

The reason for this questioning is that in this century, science has appeared to some to give primacy to technique – i.e., to its applicability. By the way, the ideology derived from technique is entirely devoted to the concepts of utility and consumption. Global society would then be no more than a rational enterprise whose objective is its own satisfaction.[18]

In this context, individual rights are perverted into a new narcissism: it is the right of a woman to have a child, the right of the dying to choose their death. At its best, is the law, which is the law of an artificial world with no limit, here to satisfy our own desires? At its worst, is it an instrument of the exploitation of a great majority by a rich minority, as is testified to by organ trafficking?

Of course, there is some reality in these arguments. The concept of individual rights is sometimes too rigid, and could amplify the prevalent individualism in our industrial democracies. The right to procreate is certainly very limited in an overcrowded country, and the ban on paying for human products would seem a dangerous nonsense when selling blood can allow a family to live.[19] This is probably true it we think about individual rights as an isolated legal concept, but as we know, individual rights can also be thought of as a political concept, founded on the constructive power of universalism.

2. Legal Universalism

For us, the idea that human rights could provoke a kind of legal selfishness is absurd.[20] Conversely, the philosophy of human rights implies universalism, which means that such rights are not the expression of an individual omnipotence, but that they symbolically draw the line between the public sphere and the private sphere. New – economic and social – rights, passed for the benefit of each individual, do not question this approach because such rights have their own limits. Finally, far from submitting to the economic sphere, the concept of human rights promotes the emergence of a common interest while separating the public and the private fields. I see another advantage, at least for those who believe in democracy and human rights, in keeping this system when thinking about the status of the human body. Human bodies are all different, while human beings – that is the power of abstraction – have equal rights.

Let me add two remarks to "prove" that legal universalism is not necessarily opposed to the knowledge brought to us by the progress of biology: First, scientific knowledge is in itself a form of universalism.[21] Secondly and consequently, it encourages us to be aware that "any regulations adopted domestically to set up limits to new techniques" have very little chance of being efficient if they are not matched by international principles to guarantee our human rights.

A final reason why I am very reluctant to see how we could get rid of the system of human rights (as the best way of protecting the human body) is that I cannot see how we could set up any other system. How is it possible to create a system that is outside the control of human reason? How is it possible to find a philosophical crossing point that serves to determine the direction to be followed? Such an approach would not be compatible with the institutions and structures of a modern society which decides by itself which rules have to be followed and therefore which rules can be modified.[22]

At this point in my paper, I would like to raise three issues: are we going to see another kind of individuality, another type of family, and/or another category of state?

B. *Reconstructing the Concepts*

1. What about Individuality?

Biomedical techniques (because they have opened up the possibility of reconstructing the human body) have raised more acutely the question of the legal nature of the human body. So, will the irreversible effect of biotechnology on the intelligence and behaviour of the individual, but also on the genetic characteristics of the human species, lead to a definition that some will call the genetic identity? Furthermore, will it lead to the protection of the individual as such or the genetic components of the individual? Until now a person could not simply be identified with the body or with the mind. A person is a legal subject, which means that in order to determine whether the human body before life or after death is a human person, we should wonder whether such a body is the substratum of a legal subject and, in my view, given the present state of the law it is difficult to give an absolute and positive answer to this question. We should therefore acknowledge that the lawyer, like the whole community, has to face new responsibilities in finding an adequate system of protection which should be respectful of human dignity without desecrating the human body too much.

2. What about the Family?

We all know that during the past decades, the sociology of the family has considerably changed: divorce has become normal, and unmarried couples and one-parent families are no longer an exception. The spread of reproductive technologies plays a particularly interesting part in this development because it focuses on the importance of the individual will to found family links. Biological and genetic parentage is no longer an absolute prerequisite for establishing such links. Of course, in the majority of cases, normal parenthood is still part of the individual parental project, but this project can often draw on the genetic or even, as in surgery, the physiological contribution of others – the donors. These contributors would not be considered by the would-be parents as having any parental rights to the unborn child, unless, for reasons of public policy, the given practice was regarded by society as illegal.

3. Finally, What about the State and Society?

Should we be afraid of the possible alliance between the existing biological mastery and the coming neurobiological one? In fact, the prospect of a different society – more repressive or more democratic – is already here through the role played by the new bioethics institutions and the internationalisation of bioethical issues. We can seriously worry about the decision-making process in bioethics: control by society seems inefficient, self-regulation is confused, and ethics bodies are sometimes too institutionalised.

It appears all too often that in the normative process – but I am speaking as a lawyer – ethics and law are confused. Therefore the risk exists that individual ethics will override what should be a social rule, valid for all members of society. But another danger exists: the danger of promoting a dogmatic moral order imposed on each of us because ethics as a transcendent phenomenon is considered superior to legal rules.

A modern state should find a balanced position between these two risks, encouraging what I will call a pedagogy promoting law and values, and giving to each component of society, individuals, professional bodies, ethics authorities, and public authorities its role and responsibility in the working out of political choices within a democratic framework and with respect for human rights. Such a new state could not ignore the necessity for international cooperation in these fields, as the present European approach aiming at elaborating a European convention on bioethics shows.

Finally, would a good methodology, an efficient policy reinforcing the rights of the individual, protecting the family and offering opportunities for social changes, be sufficient to give a legal and institutional answer to the fears raised in Society by biomedical issues and their potential impact on our human nature? Let me simply assert that such a task can only be meaningful if it reinforces our responsibility and solidarity towards others, particularly the most vulnerable among us.

University of Poitiers

NOTES

1 J. P. Kauffmann, "Mon corps...," *L'Evénement du Jeudi*, No. 455, 22 to 28 July 1993, p. 83.

2 In: C. Lombois, *La Personne, corps et âme. Journées R. Savatier*, Poitiers, 1993.

3 R. Dierkens, *Le Droit sur le corps et le cadavre de l'homme*, Paris: Masson, 1966, p. 27.

4 J. C. Galloux, *Essai de définition d'un statut juridique pour le matériel génétique*, thesis, Université de Bordeaux, 1988, Vol. 1, p. 3.

5 J. Carbonnier, "Terre et ciel dans le droit français du mariage," *Et. Ripert LGDJ*, 1950, I, p. 331.

6 J. Carbonnier, "Les personnes," *Droit civil*, Thèmes PUF, 1990, No. 4, p. 17.

7 G. Cornu, "Introduction, " *Droit civil*, Domat-Monchestien, 1988, p. 165.

[8] *Ibid.*, No. 479, p. 615.
[9] M. Harichaux, "Le Corps objet," in *Bioéthique et droit*, PUF, 1989, p. 131.
[10] Article 7–1 of the UN Convention on the Rights of Children.
[11] J. Carbonnier, op. cit., p. 19.
[12] M. T. Meulders-Klein, "*Le Droit de disposer de soi-même,*" in *Xème journée d'Études juridiques*, Bruxelles: Jean Dabin, Bruylant, 1982, p. 240.
[13] M. Harichaux, *op. cit.*, p. 130.
[14] M. A. Hermitte, "Le Corps hors du commerce, hors du marché," *Archives de philosophie du droit*, 1988, p. 338.
[15] See J. P. Kauffmann, op. cit.
[16] *L'Homme, la nature et le droit*, Paris: Ed. Bourgeois, 1988.
[17] See "Conseil d'Etat," "De l'éthique au droit," "La Documentation française," Paris, 1988.
[18] B. Edelman, "Sujet de droit et techno-science," *Archives de philosophie du droit*, 1989, p. 166.
[19] Th. Cornavin, "Théories des droits de l'homme et progrès de la biologie," *Droits*, Paris: PUF, 1985, No. 2, p. 103.
[20] L. Ferry, "L'Humanisme juridique en question, réponse de B. Edelman," *Droits*, Paris: PUF, 1991, No. 3, p. 48.
[21] G. A. Langaney, *Les Hommes: Passé, présent, conditionnel*, Paris: Colin, 1988.
[22] Manetakis, in: "Six journées juridiques franco-helléniques," *Journal de la Société de législation comparée*, 1992, p. 297.

MARTA KUDELSKA

THE HUMAN BEING – A COSMIC BEING

The following considerations are contributions to a discussion concerning the value of life and the limits of medical intervention. The basis for this thesis is a classical Hindu system, which is not only a system of values but a whole philosophy of life. To what extent does the Hindu philosophy of life, with its law of karma which somehow limits freedom of choice, allow human intervention in matters of health and sickness?

In India one recognizes a migrating subject, taking the form of either a soul or a subtle body or mind. The subject is the carrier of a karmic deposit, i.e. the sum or rather the result of good and evil deeds, intuition and thoughts. Due to individual karma, a soul in a migrating cycle, *samsara*, adopts a certain body. Samsara is treated as a state of sickness, whereas the state of health one should aspire to is the state of full liberation from karma, the state of *moksha*. The physical body is only a carrier, a tool for the migrating subject, as only the migrating subject is gifted with consciousness. Therefore, a person acts consciously thanks to this unchanging factor, wandering through succeeding incarnations in samsara. A migrating subject is conscious irrespective of whether it inhabits a body or not. The birth of a new person is the simple consequence of the act of conception, and requires three factors: the migrating subject, the mother's blood and the father's semen. It should be stressed that the factor of consciousness is decisive here, as the choice of parents is entirely dependent on karma. So, a birth in any state is not accidental but is the result of the karma accumulated through successive incarnations.

The law of karma says that each cause brings certain effects and each deed must be completed, otherwise the balance of the whole universe would be shaken. Similarly, a life in every form must be completed, and any disturbances in it are connected with very unpleasant karmic consequences.

From the medical point of view, it is very interesting to decide to what extent the condition of a person (his/her state of health (and sickness)) is determined by karma and hence must be gone through, and what the possibilities of improvement are, i.e., treatment, within these limits. The model physician would need to be equipped with adequate knowledge to decide which sicknesses are strongly influenced by karma and yet which ones can be cured.

A.-T. Tymieniecka and Z. Zalewski (eds.), Analecta Husserliana LXIV, 135–144.

In this paper I would like to put forward the Hindu concept of the person, as described by the oldest classical Hindu texts (the Vedas). The physical aspect of the matter will be presented mostly in relation to the oldest medical treatises dealing with this matter: the Treatises of Caraka and Suśruta. This means that this concept originated in the middle of the second millennium B. C. and reached its classical form presented here, in the second century A.D.

In India, philosophy and classical medicine were very closely connected with each other. When I use the term "philosophy" I mean the orthodox, brahmin systems, which are the source of Hinduism. All these systems recognize the law of karma, a cosmic, moral law, conditioning the fate of a soul in the samsaric migrating cycle. They acknowledge the fact that samsara is a state full of suffering, but also assert the existence of a state free from suffering, of full liberation from samsara and the influence of karma – a state called *moksha*. In their diagnosis of the state of the soul, Hindu philosophers directly refer to medical treatises. They compare the suffering of the soul, wandering in samsara, to human sickness. Being sick is the opposite of being cured, which corresponds to the state of *moksha*. Before prescribing a medicine, one should diagnose the sickness and find its causes. Generally speaking, in philosophy, some sort of ignorance of one's condition is always the reason. Therefore curing can be compared to enlightening a patient to the fact that his/her migrating soul is sick, and showing the way to recovery. After diagnosis and finding the causes of sickness, the right road to recovery is shown and the soul is taught how to act in order not to break the law of karma.

One more remark of a more general nature: Though the final result of Indian philosophy often seems very abstract and speculative, it attributes a special role to experience. Total liberation is the ultimate aim of all considerations. Regardless of the starting point, one aims in one direction. A practice, a mystic experience, is the very basis of all soteriological considerations. Acceptance of it is more difficult than the acceptance of a common experience, all the more so because it could only be verified by another person at the same, high level of spiritual development. The words of wise men and yogis were traditionally taken as proof of their actual experiences. A person gifted with influence was called "*apta*", meaning "one who has accomplished something". The oldest medical information, found in "Atharvaveda" and "Ayurveda", was not only vital to ancient Hindus, as it was given by *aptas*. Strictly medical information, according to the majority of Hindu medical schools, is verified by common experience.

The matter discussed here, which is a bit transcendental in nature, namely the reasons for a person's coming into being, is considered by Hindus to be certain and proven simply because it has been put forward by persons gifted with authority.

The basic assumption is that a person's birth does not originate only in our world, but is a sort of derivative from another, higher world. It is a result of samsara and the law of karma. As everyone knows, the law of karma in its cosmic aspects conditions the development of the whole world as well as of societies and individuals. The roaming of an individual is seen from a rather moral viewpoint. Each deed, thought, and intention results in the accumulation of karma, which conditions the fate of a soul after death and also affects some undertakings in the present life. Acting in the present life, we take responsibility for the life of a creature that will be the continuation of our life in the cosmic sense. The fact that most creatures do not have any recollections of their past incarnations and continuity of consciousness through consecutive incarnations does not free them from this responsibility. According to Hindus, a birth in the form of a person has an exceptional meaning in the context of immemorial wandering in the cycle of life and death, i.e. in samsara. And a person's duty is to use the precious human body in the right way, which is something most people do not realize.

Caraka, the author of a classical medical treatise, constructed a theory concerning the human body, which at the same time is related to the theory of *samkhya*. According to Caraka, the human body consists of five elements: infinities – *akaśa*, air – *vaju*, fire – *agni*, water – *apas* and earth – *prithivi*, and is in itself an abode of the sixth element, consciousness – *cetana*. The semen consists of four elements, excluding the all-penetrable *akaśa*, and mixes with it in a woman's body. To create an embryo it is not enough for the father's semen and mother's blood to unite. Such a unification may be productive only if a spirit – *atman*, together with a subtle body, consisting of *vaju*, *agni*, *apas*, *prithiwi*, and *manas*, understood as an organ of perception and thought, penetrates into this connection as a result of individual karma. The spiritual disposition of a child depends on its state of mind in a previous incarnation. So, if the previous state was divine, the child is clean and full of energy, and if the previous state was that of an animal, the child is impure and ponderous. When a person dies, his/her soul, together with a subtle body and a subtle state of *manas*, move to another womb, depending on the karma, where, in combination with a mixture of semen and blood, the conditions necessary for the origination of a new life start.

In various medical and especially philosophical texts, the subject of migration in samsara is viewed slightly differently. According to some texts, the soul, stained and burdened with karma, migrates; according to other texts a subtle state of *manas* is the carrier of the karma; according to others a whole subtle body migrates through consecutive incarnations in samsara. But it is always the migrating subject that is gifted with consciousness, whereas the physical body is secondary. These are of course very interesting problems, but what is important for our considerations is the fact that the birth of a new person is not only dependent on material factors such as the parents' bodies, but also, or mainly, on a transcendental factor, on some sort of consciousness. The relationship between these three factors: father, mother, and a conscious, migrating subject is conditioned by the impersonal, cosmic, moral law, the law of karma.

Making things simple, one may say that a person belongs to two worlds: a material one received from his/her parents, and a cosmic, spiritual one from which his/her soul comes, which conditions the functions of his/her consciousness. All but a few extremely ascetic schools of thought agree that the role of the body, though very vital, is inferior to that of the soul. The body is the soul's vehicle in the samsaric cycle, helping it on the way to *moksha*. Of course, the choice of one body and not another is not accidental; neither is the choice of one human birth form and not another – it all depends on karma. So the Indian concept is in a sense deterministic: we are shaped by previous karma and with our present conduct shape our future fate. Our fate is determined in the sense that karma, accumulated by the soul during previous incarnations, must turn into the issue of the day.

There is a theory of three supportive elements – *dhatu*, which are at the same time the three main factors in the human body. They are: *vata* (*vaju*) – the air, the wind; *pitta* – the gall; *śleshman* (*kapha*) – phlegm, lymph. *Dhatu* literally means "something that supports", so these are the main elements supporting the body. They should stay in balance, a dynamic balance, being from its nature in constant motion. Any unsettlement of this balance is pathogenic. A factor which is helpful in keeping this balance is fire, manifesting itself both as digestive fire and as the fire of asceticism – *tapas*. The basic way of maintaining a healthy and happy body is to keep the *dhatu* in harmony. This is how the texts see it; the main purpose of "Ayurveda", a leading medical textbook, is to present all sorts of diets, medicines, and the right conduct, in order to keep the *dhatu-dhatusamya* in balance.

Let us pay closer attention to a factor which is helpful in keeping this balance, namely fire. If fire appears both as ascetic and digestive, it draws attention to a specific aspect of medicine. It certainly indicates that medicine, like other disciplines in India, was subordinate to the question of redemption, the staying point for any person aiming for liberation who needed an effective tool for transforming himself at more subtle levels of consciousness.

Attributing the main role in staying healthy to some energetic factor is nothing strange. What I find interesting and very meaningful is the fact that this factor, called fire, manifesting itself as digestive fire and on a more subtle level as ascetic heat, is clearly defined as an immemorial factor, revealing itself from the very beginning of the world. So this fire, playing such a vital role in any human body, is the manifestation of immemorial, cosmic energy. A person, conscious of his/her nature's participation in cosmic energy, feels part of it and knows that all his/her actions towards the world and him/herself must become part of a general harmony. Only an action in accordance with the harmony of the universe transcends the law of karma and allows the liberation of a soul from samsara.

As I have already mentioned, birth in a human form is something unusual in the light of the cosmic wandering. It attests to positive karma accumulated in previous incarnations. The aim of the soul is to free itself from the activity of the law of karma, breaking the chain of samsara, which is possible only through the conscious transformation of consciousness. The position of a person, as presented in ancient texts, is unique and especially distinguished. I admit that the whole of Eastern thought is permeated with the idea of unity and respect for all forms of life in the whole universe; still, a human is presented as an exceptional form of existence. Wandering in the samsaric cycle, a spirit works very hard to be reborn in a human body, as only such a position gives the possibility of consciously working on oneself and reaching final liberation. Only a person has the means necessary to reach *moksha*: a subtle lively breath, the mind and ability to distinguish between real and unreal things and to consciously transform himself. The primacy of spirit over matter is the leitmotiv of the whole of Indian thought. A person concerned only with the spiritual part of his/her being was in India more esteemed than one striving for worldly well-being. This was manifested in the leading role of brahmins and priests in Hindu society up until the present. Assuming a human shape in any form has an initiative, sacred character and is the beginning of the way, but, at the same time, the beginning of the way back, to the springs. And that is why a person, by the very fact of being born a person is obliged to shape his/her spiritual nature. Otherwise, s/he may wait a very long time for an equally favourable situation in the samsaric cycle.

So birth in any human form is a positive and desirable situation. Even if this situation may seem very painful and brings unhappiness, it is unusual. From the Indian point of view, suicide is no solution at all. A person cannot escape punishment, as it is a result of previously accumulated karma which must be burned out. If a person stops his/her sufferings through suicide, his/her soul must return to the analogical situation in the future. It may wait a very long time for an equally convenient situation, as any interruptions in the activity of mature karma have very strong, negative results; suicide is often called a deadly sin. So the situation from the past will be repeated, this time in more inconvenient circumstances.

One may see the exceptionality of a person's position from the soteriological point of view; then a person seems to be the most perfect expression of existence, a creature capable of helping a soul in its escape from the activity of karma. But the uniqueness of a person is equally stressed by cosmogony, connected with the religious landscape of ancient India. The basic cosmogenic concept, though worked out in ancient times, is still widely believed in, in contemporary India and other Hindu countries. Forgetting cosmogenic concepts which are a contribution to purely metaphysical speculations, the Hindu religious system was shaped by the idea that the beginning of the world was the result of a sacrifice. Sometimes this sacrifice is presented as the oblation of the first cosmic person, like in a famous hymn about *Purusha* – "*Purushasukta*" from the tenth *Mandala* of *Rigveda*. That is why each appearance of a person on earth is understood as an imitation of the first cosmogenic activity. From this point of view, a person is a sacred "existence".

According to cosmogenic thought influencing the religious picture of India, it is clear that the original state is more perfect than the world experienced by the senses and the mind. One should seek to attain the state of boundless existence, in keeping with, and maintaining, the divine harmony. So all the actions mentioned, subordinated to the law of karma, have one purpose – not to disturb the harmony. A person's duty is to preserve this natural order and not to interfere in it. This order has a divine nature and one should enter into it. Any "corrections" of nature and any attempts aimed at breaking the influence of mature karma, e.g., any abrupt interruption of life, are not allowed. I have already mentioned the problem of suicide; the issue of unborn children has been treated similarly in India since ancient times.

A human being is of exceptional origin and the precious human body should be utilised to the maximum. A human being has an individual character from the very beginning, already in the mother's womb. In Hinduism, it is pointless to discuss when a fetus is blessed with consciousness. During intercourse, a migrating subject, already gifted with consciousness, penetrates into the parents. We may discuss the stages of the manifestation of consciousness, but for Hindus, the fact of being blessed with a separate, individual karma at the moment of conception is not debatable. As the holy texts say, a person killing an unborn child exposes himself to a deadly sin, comparable only to killing one's mother or a brahmin. The karmic consequences for such a person are very deep.

How do problems of a medical nature fit into the Hindu concept of the universe? Does the Hindu concept of non-interference in the order of the world make any treatment possible? What are the boundaries of a physician's intervention?

One should investigate this problem bearing in mind the specific role and exceptional significance of the physician in classical Hindu culture. Once more I would like to stress that all my considerations have a postulative character; I present here an ideal picture which formed and still forms a pattern of behaviour. According to holy Hindu texts, a physician should possess exceptional knowledge, comparable to a yogi's knowledge, as it is for him/her to decide when and how to intervene, when such an intervention is "harmony restoring", and when it disturbs the order. According to a classical theory, the human body is made up of elements, of both supporting (dhatu) and pathogenetic character. They are *vata* (*vaju*) – the air, the wind; *pitta* – the gall; *śleshman* (*kapha*) – phlegm, lymph. *Dhatu* literally means "something that supports", so these are the main body-supporting elements. They should be in dynamic balance. Any unsettling of this balance is pathogenic. A factor helpful in keeping this balance is fire, manifesting itself both as digestive fire and the fire of ascesis – tapas.

The main task of a physician is to keep the *dhatu* in harmony with one another. Thus, the purpose of medicine was not to look for innovative solutions, but rather to maintain the primary harmony. So any actions of a doctor aimed at improving something that had gone wrong were allowed, but only in the sense that the primary order had been distorted and needed to be restored to its ideal form. Any interference or attempts to change this order were not allowed. Ayurveda, a leading medical textbook, presents all sorts of diets, medicines, and the right conduct necessary to maintain the *dhatu-dhatusamya* balance.

From "Carakasamhita", we can learn that a physician should be devoted to the teachings of Atharvaveda, i. e. not only to medical but also to ethical and religious sciences. It is said there that a physician should know various methods of treatment, with advice concerning the rituals to be conducted and gifts to be brought. S/he should know the effects of ascetic exercises, purifying rituals, diets and magic songs. As is clear from the above advice, a physician was not only a doctor, but also played the roles of therapist and priest. S/he viewed a person as a complex entity; spiritual functions were not separated from physical ones. All medical texts, including later ones, stressed that both sickness and health do not depend only on material factors.

Herbs and minerals played a big role in treating and preventing sickness. One was also able to differentiate between the symptoms of various sicknesses, and to separate reasons from symptoms. The knowledge of suitable practices and drugs ensured not only a long, happy life, but also the attainment of the state of immortality – *brahman*. Such statements link medical knowledge with the Hindu philosophy of life, stating that all activities should lead to external, everlasting existence in a different, real state.

From the strictly medical point of view, it was acknowledged that a person belongs to two worlds: a material one inherited from physical parents, and a cosmic, spiritual one from which his/her soul comes, conditioning the functions of consciousness. And it is in this way that a doctor looked at a patient, a creature consisting of physical, changeable factors, but, at the same time, an immortal being.

While examining a patient, a physician gives a diagnosis on the basis of symptoms and after discovering the reasons, is able to prescribe a treatment. In other cases, she/he may use herbs, minerals, suitable exercises, or a diet; in others still s/he may advise conducting a ritual.

In order not to disturb the balance, one should know which levels of a person's functioning are attached to particular levels of existence. Various reasons for creating things were considered in just this context. They were: the nature of a thing – *svabhava*; god's reasons – *Iśvara*; time – *kala*; accident – *yadriccha*, fate – *niyati*, evolution – *parinama*. So the creation of a person in a mother's womb comes from the nature of the thing. And the fire which, as we said, manifests itself as digestive fire and *tapas*, is attributed to the Iśvara level. The time – kala, is the season of the year which may either help or prevent the appearances of sicknesses. Fate means virtues and vices, prices and punishments, manifesting themselves as tendencies to sickness or positive states. Later commentators stated that the five reasons, including Iśvara, are the various means of nature's self-manifestation – *prakriti*. Now we can see that in order to achieve a primary balance, one may reshape all levels of manifestation and functioning – prakriti; and a helping factor or rather the subject of activity is Iśvara, a divine factor.

The duration of a person's life in the iron epoch, *kaliyuga* (our epoch) is about one hundred years. Sinful life may shorten this span. Usually sins may reduce the length of life, manifesting, for example, as death resulting from sickness or poisoning. We are on Hindu ground and we mean deeds from previous incarnations, which formed such and not another layer of karma. But there is no determination here, and through decent actions one may reduce the possibility of premature death. Different authors have different opinions concerning certain sins or being under the influence of strong spells. But taking into consideration the most dominant point of view, supported by Yogasuta authority, one may assume that our deeds determine birth, the length of life and a nature of both nice and unpleasant sensations. In general, deeds from the previous life condition the sensations in the present one; and the present ones work for the future. Caraka says that only the results of very strong deeds are not changeable, whereas results of minor sins can be diminished by proper medication. So, in accordance with his/her karma a person may be sick, but unless it is the result of extremely evil deeds from previous incarnations, she/he may be cured by medicines.

We saw that a person participates in two levels of existence: the material level – changeable, and the spiritual level – unchangeable. Also, the reasons for creation may be reduced to material ones belonging to prakriti, and to God's reasons belonging to Iśvara. All distortions in a body's functioning have both obvious and karmic reasons. Therefore the curing process consists of first finding the nature of the cause and then, taking into consideration all inconvertible consequences, regaining the balance that is the primary harmony. If medicine is understood as treatment, correcting or looking for new ways but only in order to find the primary, healthy state, it would be allowed. But if it were to lead to experiments, attempts to impose something that was not present in the primary divine state, it would be impossible.

Looking at the picture of the person I have presented above, we can see in it some changeable, mortal elements exposed to sickness and death, but also an immortal, invariable element bound in the world of samsara as a result of a cosmic-moral law of karma. The state of a migrating soul is a morbid state, undergoing treatment. The ideal state of health, liberation, is both a primary state and the one a person should strive to attain. Sicknesses tormenting a person usually have their reasons and are curable. A physician should possess an unusual knowledge; besides the knowledge of physical conditions and common sicknesses, she/he should know what services or rituals to conduct, apart from prescribing medicines. Because sicknesses are also thought to be caused by spells and witchcraft, a physician should know how to avert them. While diagnosing a patient, she/he should have an insight into the structure of the whole reality.

Does an ideal physician resemble a yogi? This comparison is justified, especially when we look into the most primary pictures of yoga methods.

Yoga is a very ancient method, typical of the Indian continent, for transforming consciousness, with some practises even of a pre-Aryan character. If we investigate the evolution of this spiritual method, it becomes clear that at the very beginning a common aim of its practice was freedom understood not as individual liberation from samsara, but as freedom from all obstacles, which was to lead to immortality and ruling the world. A yogi practised various magic and ascetic rituals, and got to know his/her body and mind in order to subordinate reality and be able to influence its shape. And is this not one of the tasks of medicine? Is science not a contemporarily transformed type of magic thinking?

In the oldest classical texts, we see paragraphs presenting a deeply magic way of thinking. A yogi is addressed there as one who knows, who, knowing the structure of reality and its mechanisms, is able to influence it with the help of certain means. All rituals or magic spells are helpful within it. Rituals connected with birth and death were very common, the latter even more so as the right performance of them conditioned either a total liberation or, if such was not possible because of karmic reasons, a better subsequent incarnation. And one cannot overestimate the significance of the belief that subsequent births can be influenced. The texts also quote special spells and rituals which were thought to be effective in determining the sex, looks and predispositions of an unborn child.

Vyasa, the main commentator of a classical yoga textbook – "Yogasutra" of Pantanjali – explains that after reaching the ability to see reality, a yogi perceives nature, and physic, and also acquires an ability to create new manifestations of reality from certain elements, but does not do so for ethical reasons. Therefore, the main obstacle in any transformations is of an ethical nature. The law of karma takes into consideration not only a physical picture of activities but also, if not primarily, the intentions accompanying the activities. The more subtly the activity is performed, the bigger the role of intention. And the problem of transforming nature or consciousness touches a more subtle level, so any interference will play a vital role. In all classical Hindu texts it is said that an activity performed for impure, egoistic, or demonic reasons draws behind it a very strong, negative karma. The higher the level of spiritual development of a person who has performed an evil deed, the stronger the karmic consequences awaiting him/her. Such a person simultaneously loses his/her previously accumulated power. Finally, activities which are not in harmony with the world always bring undesirable results.

Modlnila

MARZENNA JAKUBCZAK

THE PHILOSOPHICAL FOUNDATIONS OF YOGA THERAPY

The word *yoga* refers to an enormous body of spiritual principles and techniques which developed in India over several millennia and which may be regarded as the very substratum of the cultural life of Indian man. Yoga is thus by no means a uniform tradition. However, it is assumed that the goal of all the forms and schools of Yoga involves the control of every process of the body-mind complex as well as the transformation of consciousness, which thereby allows consciousness to break through onto a different level of cognition.

Certain therapeutic Yoga techniques have also become popular, and are spreading throughout the Western world. In India as well as in Europe and the United States, many different schools and institutes have come into being with the aim of popularising Yoga. Not uncommonly, however, these have a sectarian and commercial character. The various methods of Yoga, known since ancient times, have often been "adjusted" to suit the contemporary needs of impatient Western Man, and have usually been simplified and reduced to a set of gymnastic or relaxing exercises devoid of their original philosophical background.

However, Yoga should not be identified solely with keep-fit exercises or even with the "pure" practice of certain consciousness-transformation techniques. The theory and practice of Yoga are not, in fact, separable categories; they mutually inform each other,[1] and because of this Yoga should be regarded as a theory-practice continuum. Thus, to apply the therapeutic techniques of Yoga properly, it is important to be aware of their philosophical foundations.

Current analysis of the philosophical presuppositions behind Yoga does not take into account all polymorphous phenomena included in the Yoga tradition. It focuses only on two types of Yoga which are the most important and representative of the Yoga tradition as a whole.

Although the first formulations of the Yoga method are to be found in the *Upaniṣads* – ancient Indian scripts originating in the fifth century B.C. – the first fully elaborated and detailed classical exposition of Yoga is contained in a treatise composed around the third century A.D. entitled *Yoga-Sūtras*. This work, attributed to Patañjali, represents the basic treatise of the classical

A.-T. Tymieniecka and Z. Zalewski (eds.), Analecta Husserliana LXIV, 145–151.

Yoga system, known as Rājayoga – one of the six schools of the Brahmanic philosophy.

A highly developed system of psychosomatic techniques is also provided by another form of Yoga known as Haṭhayoga, which is a continuation of the experimental approach of Tantrism. This type of Yoga probably originated in the eighth or ninth century A.D. Tradition celebrates the legendary Gorakṣa as the founder of Haṭhayoga. Among the later manuals of Haṭhayoga based on the works of Gorakṣa are *Śiva-Saṁhitā, Gheraṇḍa-Saṁhitā and Haṭhayoga-Pradīpikā.*

THE MAIN PRESUPPOSITIONS BEHIND YOGA

When discussing the philosophical foundations of Yoga, we must first of all keep in mind the three fundamental presuppositions common to every yogic or gnostic school in India. These are:

1. the axiom of the universality of suffering
2. the twin doctrine of rebirth and the law of moral causation
3. the doctrine of the possibility and desirability of emancipation through transcendental knowledge.

1. **Suffering** (*duḥkha*) is taken as the exact antithesis of the blissful Self (*puruṣa*), which is by definition infinite, never changing and beyond all fear and grief. But suffering does not merely imply physical pain or mental agitation. It stands for the inescapable truth of the transience of phenomenal existence.[2] Sorrow is everywhere, even in joy, for behind joy there always lies the anxiety of losing it and the fear of what may happen when such pleasure has faded. Suffering has a universal character and is a necessity of all life or being-in-the-world. The only way to escape the impact of suffering is to see things as they really are by realising one's true nature of the Self. Classical Yoga emphasises the virtue of meditative discipline as the best way to remove suffering and acquire transcendental knowledge. According to the author of *Yoga-Sūtras*, the preliminary condition for the realisation of the Self is the restriction of mental fluctuations (*citta vṛtti-nirodha*)[3] which are conceived of as a source of never-ending suffering and bondage. Normally, the Self identifies itself with the finite and limiting mind which creates all its diverse psycho-mental states. This Self's false identification or self-delusion, called "nescience" (*avidyā*), is the nourishing ground for all the various painful whirls of the mind (*kleśa*), such as egoism, attachment, aversion and thirst for life. The restrain (*nirodha*) of the psychic flux leads to the final emancipation, that is to liberation (moksa) of the Self.

In contrast, Haṭhayoga promotes "realisation through the body" and its primary intention is to remove all physical pain, restore health and prepare the body for higher spiritual practices. In one of the Haṭhayoga manuals[4] we read that the goal behind all the various methods of Haṭhayoga is the attainment of perfection through Rājayoga, i.e., classical Yoga.

Such an understanding of overwhelming suffering does not lead to pessimism because it has a positive aspect insofar as it acts as a stimulus for man to transcend it. In contrast to Christian belief, the Indians do not consider sorrow to be the result of primeval sin for which man was punished by God. Rather, the roots of sorrow lie in man's lack of Self-knowledge.

2. In classical Yoga, the underlying conception of **the doctrine of rebirth** (*saṁsāra*) and **the law of moral causation** (*karman*) consists in the conviction that each volition leaves a corresponding "impression" (*saṁsāra*) in the depth-memory which contains a network of such impressions forming distinct configurations or "traces" (*vāsana*). These are the seeds which determine the external aspects of being, such as its body or the environment in which it is born, as well as the major course of its life. The law of moral causation, *karman*, works in such a way that good deeds have as their fruit a positive after-death state, while the recompense for bad deeds is correspondingly negative. And this is why the preliminary stage of any Yoga practice involves observing both the cardinal ethical precepts and moral discipline (*yama* and *niyama*).

The depth-memory survives death and occasions new birth, thus keeping in motion the cycle of continual birth and death, called *saṁsāra*. By applying highly advanced Yoga meditation techniques, the practitioner can integrate all levels of his mind, both the presently operating empirical consciousness and the depth-memory which represents an inaccessible and unconscious layer in the normal mind-state.

3. **Emancipation** (*mokṣa*) is conceived of differently in both types of Yoga and depends on their ontological assumptions. Classical Yoga declares dualism of Nature (*prakṛti*) and the Self (*puruṣa*). When the phenomenal consciousness is "transformed" into the "witness-consciousness" representing the supreme essence of man, the emancipation or "aloneness" (*kaivalya*) of the Self is realized. Self-realisation in Rājayoga follows the subject-oriented concentration, or ecstasy (*asamprajñāta-samādhi*) and is understood in the sense of isolation from the totality of sorrowful existence.

Haṭhayoga teaches a well-balanced polarity or non-dualism (the *śiva-śakti* doctrine) and provides a means of transforming the human body into a "divine body" *(divya-śarīra)* or "diamond body" *(vajra-deha)* which alone bequeaths immortality. Thus, Haṭhayoga revives the ancient and popular ideal of "immortality" in the body which survives after emancipation in a perfected physical vehicle (*siddha-deha*).

THE BODY-MIND COMPLEX

As I have already mentioned, man is regarded in Yoga philosophy as an integrated whole. According to Patañjali, the physical, psychic and mental sides of human nature, just as in most aspects of Hindu tradition, are merely phases or modes of one and the same reality which is structured hierarchically. In *Yoga-Sūtras* all these psycho-mental organisms represent the manifestation of a single ontological objective reality (*prakṛti*), distinguished from the subjective reality of transcendental awareness that is the Self (*puruṣa*). Since the microcosm of the human being corresponds with the macrocosm, acquiring knowledge of one's own body and mind entails knowing the structure of the world as a whole.

According to Haṭhayoga, the human body is considered to be composed of gross and finer matter. Like the physical body, its subtle counterpart has a definite structure which has been the object of much practical investigation. The subtle body is precisely that part of human individuality which transmigrates and assures personal continuity after death; this means that it remains when the physical, gross body is destroyed, and lives through the lengthy cycle of births and deaths until its dissolution at the time when the living being reaches its ultimate goal of final liberation.

Although it is true that the subtle vehicle cannot claim the same "objectivity" as the physical frame, we should not simply ignore all this "subtle" physiology. Several papers have been published[5], which examine and confirm the extraordinary abilities of experienced Yoga practitioners to gain full control over the so-called autonomous nervous system, and influence both heart and breathing rhythms.

THE STRUCTURE OF THE SUBTLE BODY

The "material" from which the subtle body is built is the life-force, or cosmic or bio-energy, called *prāṇa*. Communication between the subtle body and the gross one goes on through the medium of this vital force. Despite its homogenous nature, it assumes a fivefold appearance.[6] In order to locate these appearances, they are usually associated with certain physical functions and reflexes such as breathing, excreting, circulation, coughing and digesting, etc. The flow of *prāṇa* is controlled and guided by the mind with the help of special breathing exercises, called *prāṇāyāma*. It is one of the most important Yoga techniques because the first two appearances of the life-force, i.e. *prāṇa* and *apāṇa*, are strictly linked with mental activity. When the life-force is active, the mind is also in motion. When the former is stopped, the mind also becomes tranquil.

Prāṇa is often visualized as a pulsating energy; it circulates in a vast network of channels, called *nāḍīs* which are spread out through the entire body, or rather its subtle counterpart. It is said that there are 72,000 such "channels" (but other figures are also given in the texts). Among these, *iḍā, piṅgalā and the axial suṣumṇā*, are of special significance.

Moreover, the "subtle" mystic physiology of Haṭhayoga mentions several points, called the *cakras*, which are arranged vertically along the individual's "world axis" (*suṣumṇā*), i.e. in the gross body along the spine. They are thought of as members composing the body of the serpent power (*kuṇḍalinī*) and they are frequently conceived of as "lotuses" (*padmas*) which open up as the serpent power ascends. The arousing of the latent force within the body and its controlled guidance upwards through the six centres is claimed to lead to a more complete state of enlightenment than is the case with ordinary Yoga. In the *cakras* the subtle nerves and arteries of the subtle body are connected to the physical nerves through which they receive the perception of the sense organs and communicate to the body the reactions of the subtle body and the orders of the Conscious.[7]

BETWEEN DISEASE AND HEALTH

We may consider the Yoga techniques designed to remove human suffering as a psycho-physiological form of therapy. Nevertheless, the eradication of physical and psychic sorrow or dysfunction is not the only goal of Yoga. The ultimate aim of Yoga is of a soteriological nature. It is the emancipation of the Self (*mokṣa*) and the surpassing of the boundaries of ordinary awareness through Transcendental Awareness, or Subjective Being (*puruṣa*).

The concept of illness represents a very wide category in Yoga. In fact, therapy should be applied unceasingly. This is due to three basic reasons. Firstly, even if we do not feel any pain or symptoms of illness, this does not mean that we are completely healthy (according to Yoga, even laziness or mental lethargy presage illness). Therefore, our normal state is a state of sickness, either manifest or latent. Secondly, Yoga therapy is not only thought of as a method for removing morbid symptoms but, above all, as a technique for regaining a state of health, understood in a specific sense. And thirdly, illness is so common because its causes are universal, the consequences of nescience or ignorance (*avidyā*).[8]

Nescience is the sole cause of illness. *Avidyā* is the reason for all causes of all diseases. Nescience affects both the empirical, ordinary level and its metaphysical counterpart. On the empirical level, nescience prevents us from understanding how our gross and subtle body and consciousness work. In the metaphyical dimension, *avidyā* covers up the real nature of our Self.

Thus, in order to recover health, the only solution is to get rid of the real and first cause of illness, which is the root of all others. This "recuperating" process should progress gradually. Starting with the observance of ethical precepts through purification of the body and mind, this process ought to be continued up to the moment of complete recovery equivalent to the ultimate liberation.

The absence of sorrow, or, in other words, the harmonious, undisturbed synergy of body and psyche, as believed in Yoga, does not mean that the goal of therapy has been reached. Every man, even the healthiest, has one more ailment remaining to be confronted. This is death. Thus, the purpose of Yoga therapy is not only to recover physical and psychic health, but to liberate man from the necessity of the next rebirth.

Kraków

BIBLIOGRAPHY

Avalon, A. (alias Woodroffe, J.) (1953), *The Serpent Power*. Madras.

Coster, G. (1934), *Yoga and Western Psychology*. London.

Cyboran, L. (trans.) (1986), *Klasyczna joga indyjska*. Warszawa.

Danielou, A. (1973), *Yoga. The Method of Reintegration*. London.

Eliade, M. (1984), *Joga, nieśmiertelność i wolność* [Yoga. Immortality and Freedom]. Warszawa.

Feuerstein, G. A. (1974), *The Essence of Yoga. A Contribution to the Psycho-history of Indian Civilisation*. London.

Feuerstein, G. A. (1975), *A Textbook of Yoga*. London.

Pensa, C. "On the Purification Concept in Indian Tradition with Special Regard to Yoga". *East and West*, 19 (1–2).

Sachdeva, I. P. (1978), *Yoga and Depth Psychology*. Delhi.

Sharma, S. (1995), "Terapia metod jogi a trening autogenny. Studium porównawcze" [Therapy through the Yoga Method and Autogeny Training: A Comparative Study] *Sztuka leczenia*, 1 (3).

Sinh, P. (trans.) (1992), *The Haṭhayoga-Pradīpikā.* New Delhi.
Woods, J. H. (1988), (trans.) *The Yoga-System of Patañjali.* Delhi.

NOTES

1 Feuerstein, G. (1979), *The Yoga-Sūtra of Patañjali: An Exercise in the Methodology of Textual Analysis.* New Delhi, pp. 15–16.

2 *Yoga-Sūtras* II.15: "Everything is nothing but suffering to the one who discriminates" (*Vivekin*).

3 *Yoga-Sūtras* I.2.

4 *Haṭhayoga-Pradīpikā* IV.102.

5 With regard to research on "subtle" physiology and altered states of mind being a result of yogic practices, see, for instance: B. K. Anand and G. S. China (1961), "Investigation on Yogis Claiming to Stop Their Heart Beats". *Indian Journal of Medical Research* **49**; C. S. Laubry and T. Brosse (1936), "Documents recueillis aux Indes sur les Yogins par l'enregistrement simultané du pouls, de la respiration et de l'électrocardiogramme". *Presse Médicale* **83** (Paris); G. V. M. D. Satyanarayanamurti, P. Brahmayya and M. B. B. S. Sastri (1958), "A Preliminary Scientific Investigation into Some of the Unusual Manifestations Acquired as a Result of Yogic Practices in India". *Winer Zeitschrift für Nervenheilkunde und deren Grenzgebiete*, Vienna.

6 In classical Yoga, the vital force, as *prāṇa*, extends from the nose and mouth to the heart. As *samāna*, it begins its course from the central point of the body, the navel, extending towards the borders of the body. As *apāna*, it leads down to the sole of the foot. As *udāna*, the vital force travels upwards through the vertical subtle channel as far as the head, and is especially responsible for the separation of the subtle body from the physical "vehicle" at death. *Vyāna* circulates throughout the whole body. In Haṭhayoga, a further five auxiliary forces are also mentioned.

7 Danielou, A. (1973), *Yoga. The Method of Re-Integration.* London, p. 135.

8 Patañjali distinguished five basic types of nescience (*Yoga-Sūtras* II.3.): (1) *avidyā* as misapprehension about the real nature of things, (2) egoism or "I-am-ness" (*asmitā*), (3) passion or attachment (*rāga*), (4) aversion (*dveṣa*), and (5) the will to live or self-preservation (*abhiniveśa*).

KORNEL GIBIŃSKI

TO BE (ALIVE) OR TO HAVE (GOOD HEALTH)

We began our century under the banner of heavy industry, and then there was a gradual shift towards high technology; high technology facilitated such a rapid development of natural sciences that a new watchword – the biological era – was coined. Progress in biology resulted in a better understanding of life on our planet and of the mutual dependence of ecosystems.

Furthermore, due to the avalanche of scientific progress in various fields, science itself and the technology that stems from it have started to escape from our control. At the same time we are witnessing the emergence of a new computer and information era, where communication is of paramount importance. We know that without proper inter-human communication we are threatened with a loss of bearings, subsequent dependence, and stagnation. The principle of communication, in whatever language we use, is the transfer of our ideas and thoughts. Man acts, transforming, gathering, and controlling the objects around him. These actions are described by verbs in relation to time: present, past, and future. To be, to have, and to do are the verbs describing basic human functions. There must be a reason behind the fact that the same auxiliaries exist in many European languages: *être et avoir* in French, *sein und haben* in German, *to be and to have* in English; there is a third one in English – *to do*. No wonder contemporary philosophers devote their work to the subject of being and having, to name only G. Marcel,[1] B. Staechlin,[2] or E. Fromm.[3] This subject was also used by Pope John Paul II in his encyclic *Redemptor Hominis*[4] where he discusses the dilemma of "to be more or to have more" in order to condemn the consumer lifestyle. In 1976, during a WHO conference in Geneva, I heard the representative of Togo argue that in his country medicines were not a problem of having but of being, i.e., of surviving.[5]

Used on different occasions and applied to different problems, the two words always refer to the most basic situations. As the title of the present symposium is "Philosophy, Science, and Medicine", I would like to address the dilemma of "to be and to have": to be alive or to have good health. The question is not asked with deep philosophical probing in mind. I see it as important in a more direct and practical way: it is a problem of free choice, like the choice of lifestyle from asceticism to consumerism and hedonism.

Hippocrates, known as the father of medicine, stated that the cause of disease should first be sought, followed by a rational treatment; he also stated

A.-T. Tymieniecka and Z. Zalewski (eds.), Analecta Husserliana LXIV, 153–159.

that an ethical attitude should be adopted towards the patient. For a long time it was assumed that his writings described the state of medical art in ancient Greece. Recent analysis[6] shows, however, that Hippocrates objected to purely symptomatic and palliative therapeutics and the common charlatan practices of his time. His rational directives for progress in medicine (based on a very low level of technical and scientific diagnostic methods) had to wait one-and-a-half thousand years until they were better expressed in the Cartesian comparison of man's state of health and disease to the workings of a clock. From that time onwards medical sciences began to develop quickly. The impressive beginnings of the great industrial era promised unlimited progress and unlimited possibilities in the field of the human intellect and production, leading to the full control of nature. The same hope was expressed in medical sciences. A magnificent example of such faith and achievement was microbiology and the treatment of infectious diseases at the turn of the century. The next milestones in this direction were the discovery of antibiotics and the progress in virology and immunology in the fifties, which seemed to promise total eradication of infectious diseases. Today we know that it was an unattainable absurdity and that eradication of the microbial world would have meant the end of terrestrial life. Life proved to be stronger, microbes knew how to become resistant, new species appeared on the world scene, and life continued.

Structural, biochemical, and physiological research was carried out in depth with the same hope, and we have since admired the achievements which lead us to molecular biology and a molecular understanding of pathology. Yet, recently, physicists have presented their new theory of chaos with its practical implications that the tiniest non-linear processes make the long-term predictability of complex, macroscopically observable phenomena impossible. It should be emphasized that reductionist research has worked on the assumption that once we know all the elementary components of biological processes, we shall gain full control over life. Reductionist research has spread all over the world with great impetus and will certainly continue to bring new successful discoveries for decades. But its end-point remains in question.[7]

We can't help admiring the successes of the human intellect and of technology, which have permitted us to forecast several years in advance the year, day, and hour of the announcement of the final result on the human genome. It is considered to be the top achievement in the cognition of the human body. We have already learned that on top of chromosomes, the thin organelles contained in the nucleus of every cell, there are so-called telemeres containing information on how many times the cell can divide. Due to this property of encoded proliferation, the renewal of basic elements is ensured. The worn-out cells can be replaced, and consequently human machinery is able to function for decades. Finally, with the exhaustion of the possibilities of proliferation, cells must vanish, resulting in the atrophy of tissue, organ, system, and finally in the most natural death. The fact that the longest possible life span is programmed in this way means that any campaign for the prolongation of human life beyond this threshold is nonsense, like the repeatedly proposed designs for a perpetuum mobile. Of course, man may not live for the duration of his programmed life for many reasons: a fatal accident, a defect in his DNA as a result of an inherited disease, extreme environmental temperature, toxins (bacterial, industrial, herbal, medicinal) may all be a cause of shorter life. Our task is to combat these dangers, and this means a campaign against disease and an attempt to avoid the threat, rather than a campaign for life prolongation.

Yet, our health policy traditionally operates on the indices of death rather than disease. The death rate for a particular disease is the percentage of those who die (out of all those who contract that disease), e.g., 20% in myocardial infarction, 90% in lung cancer, while the mortality index is calculated per one hundred thousand inhabitants. Another index, the so-called statistical life expectancy, also has its end-point at death. We are also firmly engaged in fighting prenatal mortality. So, we cannot say that all these indices are wrong or useless. They are the best we have. The time between birth and death can be measured precisely enough, but the diagnosis of a disease is burdened with a large margin of uncertainty, and *post mortem* verification is very rarely carried out.

However, the best indices of mortality or of the death rate do not tell us about the health level of the society. Morbidity indices, apart from defects and errors in diagnosis, are unreliable because of deficient reporting of sickness (patients avoid doctors for many reasons), and because of inconsistent cooperation in therapy on the part of patients. The latter is linked to the level of consciousness and the system of values in a given society.

This is not to say that the widespread application of mortality indices is bad. Its application in medical research and studies on communal hygiene, etc., has led to a considerable increase in the average life span. On the other hand, the culture of death negation and the atmosphere of death rejection have encouraged researchers to continue efforts and programmes in this direction. It brings them satisfaction, the applause of the media, and popular recognition, followed by priority in grant allocation for such projects.

Nobody has ever asked society whether it sees any real profit from such spectacular achievements, or whether they meet the needs of the society not only in terminal cases but also in everyday health problems. People like sensations and hits if they do not actually have any urgent health problems. But if they cannot find the expected help with their everyday problems, they force such legal regulations through their parliamentary representatives, or promote alternative medicine. Embittered, they claim that "science is a failure". Prevention in particular is mistrusted, as its effects are always delayed to a vague future and are therefore less 'tangible' than those of a potent drug or surgery. It is very difficult, if possible at all, to promote social acceptance of the idea of health promotion.

Whether people will use up the whole health capital inscribed on their genomes largely depends on their adaptive and immunological mechanisms, but also on individual actions and collective activities undertaken to protect themselves against the attacks of their environment. Protection also depends on personal and communal hygiene, and this is a matter of choice. The logic of this seems clear and the primacy of prevention can hardly be rejected; nevertheless, prevention is usually pushed aside in practice.

There are 4 million pensioners in Poland, and yet the mean age is still not as high as in highly developed countries – and the number of old people is still growing. What happens to them at the end of their life? I am not thinking here of social problems like the scarcity of nursing homes and shelters for the homeless. I am thinking of the terminally ill whom well-equipped hospitals do not want to keep because they need beds for patients who could be treated successfully. Also, local hospitals do not want to admit them, arguing that they cannot help. Families often feel even more helpless; sometimes they are equally old and disabled, sometimes they cannot take care of the sick because they all work, and often there is no room in crowded apartments. The problem cannot be solved by building new hospitals for the terminally ill, and geriatric wards are not the proper answer either. The terminally ill should not suffer, should not be left alone, and should be freely provided with analgesic and relieving drugs. A form of organized social assistance is needed to take good care of these people. This, of course, is not an original proposition. The movement for palliative medicine, simple, cheap and commonly accessible, otherwise called the hospice movement, has spread everywhere in the world. It has also reached Poland and is developing vigorously.[8]

A society which is convinced of and dedicated to an idea is able to realize big projects. Good examples are: the idea of a thousand schools for the millennium of the Polish state – a project which was to meet the needs of the demographic explosion after the Second World War – and the campaign to reduce prenatal mortality. The latter was based on the preventive care of mother and child before, and just after, birth. Its implementation depended on an intensive information campaign to make society understand the value of preventive care, as well as on the creation of organisational forms within which primary and secondary preventive measures could be executed.

What is the difference between the right to health and comfort of the newborn and the right to comfort in a dying man's last days? We look forward to the birth of a child with pleasant tension and delight, although we know that birth is connected with suffering and uncertainty. Why do we look with fear at the other end of human existence? Individual life is only a small link in the endless chain of the life of humanity lasting for hundreds of thousands of years. The departure should not be a surprise; it does not have to be painful or cruel. It often happens that the drama of death is much more acutely felt by the family than by the dying person. Tanathophoby, although a commonly dominating feeling, should be fought against. When facing a real terminal state, it is necessary to secure warmhearted care and comfort to help reduce pain and suffering in the last weeks. The progress in analgesics and other auxiliary drugs is huge, and the goal is attainable. It is a paradox and an optimistic sign that in our heavily consumer-oriented times there are so many volunteers, mostly young people, ready to help free of charge. The costs of organizing palliative care are incomparably smaller than those of heroic campaigns against death.

A new understanding of death must be the goal of a society-wide movement; a new attitude is imperative amongst both the healthy and the ill. Death must be de-demonized; the notorious skeleton with a scythe must disappear. Nobody, including children, should have a fear of death instilled in them. We should all be able to think and talk about death as an inevitable stage in life. Everything *may* happen in human life; death *must* happen to everybody. So, instead of being perceived as a painful and cruel experience, it should be something naturally expected. Such an attitude to death is possible in all countries, no matter how rich or poor they are.[9] Getting rid of tanathophoby should lead to a change in our research priorities, and thus ensure better funding of research in accordance with social needs as they arise.

As well as changing our attitude toward death, we need to abandon the habit of concealing the real diagnosis and deceiving the patient. This habit could have been justified in times when we did not know about the encoded limit of life and when death was seen as a sinister fate or punishment. I realize that the disclosure of truth to a patient may be a trauma and that a doctor may see the patient's eyes filled with fear. This has to be taken into account. The doctor must have tact, he must learn how to handle the situation. This new attitude toward death is not yet as common in Poland as it has recently become, e.g., in the West. Yet people have the right to know and this right has to be respected, as does the right to any reliable information.[10] Society is beginning to demand true information. We are entering the computer era.

Changes in human attitudes are deep, and we doctors have to be aware of them. Even if the splendid achievements of science and technology often raise our hopes and seem to fortify the culture of death rejection, we must never lose the vision of man and health policy as seen from the humanistic perspective: social, economic, and, above all, philosophical.

If we succeed in turning human thought away from death's cruelty, and from the option of longevity, in order to steer it towards energetic and persistent health protection through (disease) prevention, towards the sanctity of life in good health and with full human dignity, perhaps then it will be easier to direct the main stream of research and scientific effort towards better prevention and ensuring physical health and psychic comfort than it is during the present campaign against death. This would be to postulate enjoyment of one's natural potential for health rather than to crave longer life at any price (at a miserable health level). In other words, quality is more important than quantity of life.

It is extremely difficult to change human attitudes. This will require enormous educational efforts. Perhaps the computer era will help us to meet the challenge. Information can now be disseminated unbelievably quickly and the process is getting faster and faster. Four thousand years ago, information was communicated by word of mouth – the written word on papyrus was reserved for the privileged few. The first printing presses in Europe in the fifteenth century sped up communication by spreading the printed word (yet a hundred years later a Renaissance Italian leader confessed he was ashamed to own a printed book). At present the gigantic flow of the printed word is proving insufficient. Shares on the electronic market have reached three trillion dollars a day, exceeding the yearly U.S. budget more than twice. In 1971, a microprocessor was able to perform 60,000 calculations; in 1995 a microprocessor can do 250,000. In 1971, only 130 calls could be cabled under the Atlantic, now the number is 500,000 calls transmitted simultaneously. The question is how to use this fantastic opportunity to communicate, how to make impersonal information persuade human beings to change their traditional attitudes, and how to help them live better lives.

Slaskiej Akademii Medycznej, Katowice

NOTES

[1] G. Marcel, *Etre et avoir*, Montagne: Aubier, 1962.
[2] B. Staechlin, *Haben und Sein*, Zürich: 1969.
[3] E. Fromm, *To Have or to be*, Hasper and Celius; 1976.
[4] John Paul II, *Redemptor hominis*, Typis Polyglottis Vaticanis, 1979.
[5] T. Johnson-Romuald, "Un problème essentiellement politique", in: Z. Bankowski & J. F. Dunne (eds.), *Trends and Prospects in Drug Research and Development*, Geneva: CIOMS, 1977.
[6] N. M. de S. Cameron, "The Seamless Dress of Hippocratic Medicine", *Ethics and Medicine* 1991, pp. 43–50.
[7] Gleick, *Choas. Making a New Science*, Viking Penguin, 1987.
[8] K. Gibiński (ed.), *Terminalnie chory-Hospicjum* (Palliative Medicine), Kraków: PAU, 1996.
[9] Ibid.
[10] K. Gibiński (ed.), *Tajemnica lekarska* (Medical Confidence), Kraków: PAU, 1994.

KAZIMIERZ SZEWCZYK

(WO)MAN AND DEATH IN THE SECOND HALF OF THE TWENTIETH CENTURY

I. WORSHIP

Modernity marks the end of a period which may be called the era of faith. Ancient Greece believed in the reality of Truth, Goodness and Beauty. But above it all it believed in a human Mind that was able to grasp this trinity. Christianity believes in God, who takes an interest in human beings and their deeds. Modernity, in turn, believes in Technology.

The human being of the era of faith looked upwards; s/he was – to quote Hans Jonas[1] – "vertically oriented". There, high above, s/he sought a universal principle that would justify human existence. And, since life was transient, s/he looked up, trying to justify death, to give it a meaning and to find a remedy for fears connected with death. For the Greek or Roman thinker, only philosophy could be a remedy for thanatic fears. This was particularly visible in the work of writers who possessed a strong awareness of death, e.g. Plato, Epicurus, and Marcus Aurelius. Marcus Aurelius recommended that one should "in all situations await death without fear, regardless of the circumstances". This attitude resulted from his unshaken belief in the rationality of the laws of nature. And since these laws are rational, the death which follows them "cannot be evil".

Early Christianity had a similar attitude toward death. However, belief in nature was replaced here by belief and trust in God. For the early Christians death could not be evil, since it was the moment preceding the long-awaited meeting with the Good Creator.

But at some stage the fear of eternal damnation sneaked – probably under the influence of the teachings of St. Augustine – into this quiet and good-spirited awaiting of death. This fear of damnation grew through the centuries and as Delumeau says,[2] it transformed the natural fear of death into a "religious fear of the Last Judgement". This fear resulted from belief in God's justice and from the awareness of being sinful. It was further intensified by external circumstances: famine, wars, fire and plagues.

When experiencing these disasters, the human being does not really fear death, but Judgement and hell. Death still remains, as it did for the early Christians, as the gateway to the beyond. This gateway is created by a belief

A.-T. Tymieniecka and Z. Zalewski (eds.), Analecta Husserliana LXIV, 161–171.

in God. Faith makes death meaningful and does not leave the human being feeling totally helpless in the face of his/her finite nature and of the eschatological fears haunting him/her. The human being may leave everything to God, trusting in his wisdom, goodness and justice, as emphasised by Luther. Alternatively, despite being sinful, a human being can strive, by his exemplary life and deeds, to become worthy of redemption. The second strategy was adopted by the Catholic Church.

In the middle of the eighteenth century the process of secularisation of our culture began. It was preceded by the disappearance of the belief in hell, a punishing God and the Last Judgement. Aries describes this phenomenon as the "weakening of the eschatological idea of Christianity".[3] However, the eschatological fear which built up in the course of centuries became a "socio-psychological entity" and did not automatically vanish with the weakening of the eschatological concept. Therefore, it needed new ground for its existence and found it in death itself. Death came to be viewed as a natural/medical phenomenon.

However, nature was no longer seen as a rational force, as it had been by Marcus Aurelius (because it conformed to rational laws), but as a chaotic and dynamic force, hostile to any form of life. Fearsome death is an exit into an area (fearsome as well) of non-being, devoid of any positive meaning.

The fear of death as an exit into axiological nothingness gradually took the form of a nearly hysterical fear of being buried alive, which was present almost until the end of the last century. This fear started to disappear as belief in technology and science and its priests – scientists – increased.

The main aim of technology and science was and is to subdue nature, to bring it under human control. This science "entered" medicine thanks to Louis Pasteur in 1857. From that time onwards disease was approached in the same way as other phenomena: its cause could be discovered, and through experimentation a cure could be found for it, and thus it could be controlled.

With his discoveries, Pasteur created the hope of conquering all diseases, even fatal ones. Fearsome death could then hide behind a disease. In this way modernity destroyed mortality, to use Zygmunt Bauman's words. Dying was no longer the inevitable fate of humanity, but dissolved in a multitude of causes, which were avoidable thanks to science. "Eschatology has been successfully dissolved in technology",[4] and the previous model of *tame death* degenerated into *wild death*, rejected from the individual and social conscience.

Faith in Technology, however, did not make death meaningful, as was the case in times of faith in Mind and God. On the contrary, it created the hope of achieving something which Louis-Vincent Thomas called "technological non-mortality,"[5] infinite prolonging of life by means of medical techniques.

Wild death escapes sense-creating activities and thus becomes a very powerful source of thanatic fears. In turn, these fears feed faith in Technology, since only Technology can – due to the promise of everlasting life – counterbalance the fear of death. In modern medicine faith in Technology manifests to the full in so-called hospital ideology, particularly characteristic of the fifth and sixth decades of our century.

According to this ideology, death is the worst thing that can happen to a human being. Therefore, the struggle with disease becomes more important for the hospital ideologists than taking care of patients. In reality it is a fight between science and death, with doctors acting as middlemen. The persistence of this action proves how strong the modern person's faith in Technology and fear of his own finiteness are. This faith and this fear hasten the process of turning doctors into experts, priests of science, who celebrate their mysteries in hospitals – modern shrines of Hi-Tech.

At the end of the era of faith, medicine completely took over health, disease and death. As a result, totally "medicinized" wild death has taken the form of *awry death.* Surrounded by a conspiracy of silence, patients die in solitude, under the absolute, almost paternalistic control of doctors who deny the obvious fact that their patients do die.[6] Faith in Technology seems to relieve the modern person of the responsibility for his/her own health and disease. But the price of this relief is the deprivation of individuals and societies of their right to die.

II. AUTONOMY

One of the effects of modern human actions is to deprive reality of all meaning, making it, as Jacob von Uexküll says, a "blind and idiotic creature",[7] and thus throwing the world into axiological nothingness. It is an almost total "deconstruction of the world". Scientific and technological achievements flatten reality, destroying its vertical, axiological dimension. They also disturb the ontological solidity of the world. Reality "assumes [...] a constitution which has so far been restricted to the sphere of fine arts, the constitution of an entity capable of being produced, volatile, unbound, ephemeral [...]".[8]

Postmodernity has taken over a world from its predecessor which only very weakly resists human manipulation. In such a world the metaphors of a fight lose their relevance and are replaced by the metaphors of creating.

Likewise, faith in Technology is no longer the highest principle. But in order to create in a reality devoid of any meaning, one must first "create" meaning itself. In turn, in a world where faith in some distant transcendental source of meaning is so weak, the creator of senses can only be an autonomous human individual.

The modern person can be pictured as a train driver, driving a locomotive (the powerful forces of the world) along fixed tracks towards a target station, the existence of which this train driver firmly believes in. A postmodern individual, however, can be portrayed as a computer programmer. He creates – usually in absolute solitude – newer and newer simulated worlds on his screen. Their life span and form depend only on the programmer, who in his actions possesses the freedom of an artist.

The metaphor of the computer programmer suggests that creating meaning is, to a large extent, identical with creating a reality in which this invented meaning can function. Thus, my life is meaningful only insofar as I have pre-planned it and will follow this meticulous plan. Likewise, my death becomes meaningful when I rationally think it over and "perform" – when I have decided that the right time has come. However, in order to plan his/her own death, a human being must first be given back his/her right to die. This process started in the U.S.A. in the late 60's and early 70's. It has spread to western European countries, assuming a dual form:

1. Bringing death out of hospitals and from under the supervision of its ideologists to homes and hospices. The latter, it is supposed,[9] are better than homes since they can create proper conditions for a "peaceful death", as Daniel Callahan puts it.[10]
2. Handing over control of death to the individual. Let us consider the second form first. I will return to the hospice movement at the end of my presentation.

The first proposal worth noting, as regards putting death under the autonomous rule of the patient, was put forward in 1976 by Robert Veatch. According to this proposal, a patient should be given the right to "choose his/her own definition of death, depending on his/her own religious and philosophical beliefs". Seventeen years later, Veatch rephrased his proposal, offering three definitions of death: death of the brain as a whole, death of the heart/irreversible circulatory arrest, and a definition "focused on the cortical centres in the brains", based on the loss of consciousness, even in the presence of self-sustained breathing and circulation.[11]

According to Veatch's suggestion, a person acquires some possibility of defining what s/he considers death to be. Since, finally, as Veatch states, the problem of death resolves itself into the common-sense pragmatic definition of it, then the choice of one of these definitions by an individual is reasonable as well.

The apogee of "making" one's own death is observed in the publications of the Hemlock Society, founded in California in 1980 by the journalist Derek Humphry. These publications propagate the model of "mortally ill rational suicide", which might be called *etiquette death.* It is presented in a particularly clear way in the Society founder's book *Final Exit.*[12] The most characteristic features of etiquette death are the following:

1. Careful planning of one's death. Among others, this includes: seeking and choosing a doctor who would be willing to cooperate in preparing for death, drawing up and signing the documents that protect the doctor and the family against being accused of homicide, buying the "deadly dose" of appropriate medicine, storing it in proper conditions, protecting it from children and not exceeding the expiration date. The "Rational Activity Plan" aims to ensure:
2. The infallibility, aestheticism and "decency" of the suicide. To die decently means – for members of the Society – to pass away in an elegant, silent and painless way with the family at one's side, after a farewell evening party "with champagne and caviar served"; it means not making problems for anybody with one's decision to leave.
3. Involving other people in preparing one's death. The physician, the lawyer, and relatives become almost solely auxiliary and aesthetic elements of the rational plan.
4. The continuous presence of death in the consciousness of the individual, and the continuous interest in it. Members of the Society – the incurably ill, those close to the final exit, and those who are entirely healthy – meet in "support groups". During these meetings they share their feelings about death and dying as well as efficient ways of "carrying it out". They also develop new techniques to increase the effectiveness and elegance of a suicide, and practice those techniques.

From the description of rational suicide above it appears that for its supporters the meaning of death is identified with "carrying it out". Thus, for Humphry, it is most important to carry it out in a beautiful and efficient way. However, he is less interested in the question of how he should carry out his own death. In thanatology, just as in all post-modern culture, ethics gives way to aesthetics and praxiology. Our belief in Technology, having levelled the vertical-axiological dimension of reality, has made us nomads of meaning. We are not, as we used to be in the era of faith in God, pilgrims travelling towards a clearly defined aim, but nomads wandering among transitory oases of meanings created by us.

An analysis of the etiquette of procedures connected to death also shows that it is the cry of an individual for communion with other people. Planning a suicide enables us – through the mediation of death – to get in contact with others, to break our isolation for at least a moment. The post-modern nomad pays a very high price for his/her freedom – the price of being lonely. That is why the meticulousness which is characteristic of members of the Hemlock Society in planning their suicides is a derivative of loneliness and fear of death.

Etiquette death is a very enduring trend gravitating towards voluntary euthanasia. Judging by the name of this kind of death, agreement to euthanasia means agreement to handing over (the right to) death to a patient. However, in reality, euthanasia is, to a large extent, a repeated handing over of death to a physician; thus it is its repeated medicalisation. This is because euthanasia, at least in Holland, is acknowledged as voluntary when:

a) A patient requests it.
b) This request is submitted after the patient has been fully informed and after the request has been considered profoundly.
c) The patient's wish is durable.
d) Suffering is unbearable and without any hope of relief, and
e) An attending doctor arrives at the decision after s/he has consulted with other physicians.[13]

It is easy to see that the last two conditions hand over the responsibility for death to doctors. It is they, however – and not the patient – who, being fully acquainted with the methods of palliative care, decide whether the kind of suffering in question holds any hope of relief. It is also they – and not the patient – who, after consulting together in their own circle, arrive at the ultimate decision. Thus, voluntary euthanasia is voluntary only to a certain degree, and it might be called, in the words of Richard Fenigsen,[14] "*agreed upon death*" applied by specialists after the prior request of the person concerned.

If, after agreement, physicians acknowledge this demand to be justified from their medical point of view, "then" – as Christian Barnard[15] claims – "they are fully competent to prescribe a suicide in this case just as they prescribe taking particular medicaments in other cases". And thus the "agreed upon death" becomes the medicament chosen to cure the basic complaint tormenting post-modern people, namely the fear of being unable to give meaning to their own lives.

Consequently, life, which I am unable to make, sinks into axiological-ontological nothingness and I am transformed into a kind of a zombie – a living dead person unable to create oases of meaning and to wander among them.

Taking into account the above considerations, I think I can venture the opinion that postmodernity is at least as afraid of death as modernity. Nevertheless, the post-modern person does not deny the fact of his/her passing away as passionately, and s/he tries to decrease the fear that it brings about by, among other things, forestalling death with euthanasia or prescribed suicide carried out by specialists in a proficient way. Thus, as a medicine, "agreed upon death" is – to express it paradoxically – a demand for death because of a great fear of it. Intensive thanatic fears accompany not only awry death, but also etiquette death and "agreed upon death". In my opinion, we may consider them all as subtypes of wild death.

III. RESPONSIBILITY

In 80's Western culture, what might be called an awareness of deficit gained popularity. It concerned both philosophy and economy. The post-modern nomad became tired of wandering among the oases of meaning. Creating transitory sensations lost its meaning, and this gave rise to the feeling of their deficit. In the situation of deficit of meaning, loneliness became an exorbitant price for individual autonomy. Therefore, the issue of the deficit situation proposed by various authors was usually connected with an attempt to find an exit from loneliness.

For Emmanuel Lévinas, his responsibility towards other people is the exit. It is more original than freedom and it gives sense to his existence as well as constituting "a genuine relationship with another group".[16] The second one for Lévinas is the "face" that "requests and orders: Do not kill!" In Hans Jonas' work,[17] the whole of nature, whose existence is endangered by us – the human race that crowns it – speaks to us with the equivalent of such a request/order.

For Jonas, an awareness of the deficit is an awareness of the deficit of the ecological balance ("symbiosis") between Homo sapiens and other species of plants and animals. The evolutional and cultural development of (wo)man, having endowed him/her with power which is too great in comparison with the self-regulating environment, has led to a disturbance of the balance. Thus, the control over the world gained by people, which is equivalent to the control of a parent over a child, has to be balanced by responsibility equivalent to parental responsibility.

The nomad of meaning, first relishing his autonomy, then slightly bored with it, becomes the protector of nature, responsible for the future fate of the human race, perturbed by his/her own power and tormented by the feeling of "impersonal guilt" before future generations. This responsibility gives meaning to his/her life and – together with his/her fear and guilt – unites him/her with others in the community of those undertaking the task.

In medicine, the development of the technological power of Homo sapiens has also led to a deficit situation. Medical technology is becoming more and more expensive, but still does not guarantee a decrease in the mortality rate comparable to the outlay. So there appears to be an absolute necessity for more economical and rational managing of funds that are allocated to medicine.

Replacing the medical "model of disease" with the "model of health" in the late 80's, to use the nomenclature of Henk ten Have,[18] was a very important result of the deficit situation in medicine. The first model was worked out on the basis of the modernist belief in Technology. It took away the responsibility for health and disease from the patient and passed it on to medical specialists. However, the "model of health" makes everybody responsible for his/her own health and sickness. Maintaining health in this model depends, first of all, on the choice, made by the individual, of a lifestyle recognised by medicine as a healthy one. By living in a healthy way, we save the relatively meager funds given to the health service and, at the same time, we behave responsibly towards other people who use those funds as well.

The ability to distinguish between the real needs of potential patients and their whims is another, extremely important result of the awareness of the deficit situation. The real needs should be, as far as possible, fully provided for, but the whims should not. The deficit situation puts all of us under an obligation to give up own wants and take on responsibility for the self-realisation of other members of society.

The moral person of the last decade of the twentieth century is the individual motivated by responsibility towards other people, cultivating the virtues of moderation and restraint. The terminally ill patient should also take into account his/her responsibility towards other people when deciding about his/her death. According to John Hardwig,[19] if the terminally ill's will to live becomes too heavy a burden for others, particularly friends and relatives who are emotionally, financially and morally involved in the process of dying, then it becomes a whim. That is when it becomes morally appropriate and responsible to interrupt the life-prolonging treatment.

The approach to death outlined by the above-mentioned author might be called *negotiated death*. The decision of a dying person must be preceded by discussion with the physician and the closest relatives. Through negotiation, the question of whether life prolongation is a need or merely an irresponsible wish, must be settled.

Moreover, every person has a duty to give appropriate instructions in advance, in which s/he authorises the cessation of life-prolonging treatment the moment it becomes irresponsible acting (because in the future s/he him/herself may not able to decide consciously due to his/her lack of competence caused by disease).

The negotiated death may become the option for euthanasia on one condition: that death itself is recognised as an integral part of life by the person in the deficit situation. It requires the rejection of *passage thanatology* that has existed hitherto, in favour of *threshold thanatology*.[20]

In passage thanatology, death is either a gateway to the beyond, or an exit into nothingness in the same way as it is in the wild death model. The exit into nothingness evokes terror which surpasses the fear of the Last Judgement. This terror is very easily linked with fear and the feeling of guilt connected with the responsibility of an individual for his/her own death before relatives. In the "agreed upon death" of the deficit awareness period, the fear of nothingness is disguised by responsible concern and anxiety about the future of relatives who are threatened by the death in different ways.

In threshold thanatology, death is the last stage of life, its impassable threshold, a kind of mirror placed at the end of an individual's path. Thus, it makes a person look back towards his/her life and not ahead towards the other side; it shines with the glitter of meaning borrowed from life, if the life makes sense, and also illuminates life with its glitter. Threshold thanatology does not deny the existence of the other side; it simply does not mention the other side, leaving it to the eschatologists.

Acknowledging death as the last stage of life cannot make us entirely free from thanatic fears. However, it can diminish them to such a degree as to enable us to reject the wild death model and reduce the "overfeeling of guilt" tormenting us. Reintroducing death to life will also make it possible to transform the responsible decision to cease life-prolonging treatments into a decision concerning life rather than death. For, because of this decision, there is a chance that death will become the last stage of life's "peregrinations" – a chance to transform wild death into *peaceful death*.

Peaceful death may be viewed as a form of negotiated death if it is:

a) Under the medical attention of specialists – physicians, nurses, psychologists.
b) Among relatives.
c) A death where the fact of dying is not denied by specialists, the dying person and his/her relatives.

In peaceful death, the thanatic fears of persons involved are relieved, to a large extent, in the solicitude of staying by the side of the ill person. In addition, being there for the dying person helps him/her to overcome his/her fears. It is also a kind of reward for the patient's decision, and the decision itself gives him/her the feeling of power and adds meaning to his/her death. For it was s/he who, having chosen death, was strong enough to take the responsibility for the fate of his/her family and friends upon him/herself – fellow creatures who will pass away and need care in the same way s/he does.

The recognition of death as a stage of life by medicine puts physicians under a moral obligation to secure a peaceful death for a dying person. Thus, peaceful death becomes – in the same manner as health, life or prevention – one of the main values/aims of medicine. Together with peaceful death, preventing terminal suffering and the relentless struggle against it become principal values as well. Palliative specialists, besides staying with a patient, should also secure a death without pain for him/her. The awareness of this obligation will undoubtedly soothe a part of the thanatic fears resulting from the fear of suffering.

From the description of the features of peaceful death, it appears that the hospice, being an institution situated "in between" hospital and home, is the most appropriate place for it. The hospice should combine specialist care, which so far has been offered only by hospitals, with the intimacy of a home. The hospice, as a place of peaceful dying, teaches us to establish true relationships with others, with responsibility for them – individuals as mortal as ourselves. Therefore, for a person in a deficit situation, they may become a school of responsibility for him/her, for his/her relatives and – indirectly – for the future of humankind. Surely serving an apprenticeship in hospices will precipitate a metamorphosis of the airy nomad of meaning into a responsible guardian of fragile and endangered life.

Medical University of Lodz

NOTES

[1] H. Jonas, *The Imperative of Responsibility*, Chicago, 1984.

[2] J. Delumeau, *Le Péché et la peur. La culpabilisation en Occident XIIIe-XVIIIe siècles*, Paris, 1980.

[3] P. Aries, *L'Homme devant la mort*, Paris, 1977.

[4] Z. Bauman, *Mortality, Immortality and Other Life Strategies*, Cambridge, 1992, p. 141.

[5] L.-V. Thomas, *Le Cadavre. De la biologie à l'anthropologie*, Bruxelles, 1980.

[6] E. Kubler-Ross, *On Death and Dying*, New York, 1969.

[7] J. Uexkull, *Der Sinn des Lebens*, Stuttgart, 1977, p. 76.

[8] W. Welsh, "Asthetisierungsprozesse Phänomene, Unterscheidungen, Perspektiven", *Deutsch Zeitschrift für Philosophie*, 1993, No. **1**, p. 10.

[9] "The Technological Tether. An Introduction to Ethical and Social Issues in High-Tech Home Care, Special Supplement", *Hastings Center Report*, 1994, No. **5**.

[10] D. Callahan, "Pursuing a Peaceful Death", *Hastings Center Report*, 1994, No. **4**, p. 33.

[11] R. Veatch, "The Impending Collapse of the Whole – Brain Definition of Death", *Hastings Center Report*, 1993, No. **4**, p. 18.

[12] D. Humphry, *Final Exit. The Practicalities of Self-Deliverance and Assisted Suicide for the Dying*, Eugene, 1991.

[13] J. J. M. van Delon, L. Pijnenberg and P. J. van der Maas, "The Remmelink Study, Two Years Later", *Hastings Center Report*, 1993, No. **6**, p. 25.

[14] R. Fenigsen, *Euthanasia – The Choice of Death?* Poznań, 1994.

[15] C. Barnard, *Good Life, Good Death*, Kapsztat, 1980.

[16] E. Levinas, *Ethique et infinité*, Paris, 1982.

[17] H. Jonas, op. cit.

[18] H. ten Have and M. Loughlin, "Responsibilities and Rationalities: Should the Patient Be Blamed?" in: *Health Care Analysis*, 1994, Vol. **2**, p. 122.

[19] J. Hardwig, "Support and Invisible Family, Dying Well in the Hospital. The Lessons of Support, Special Supplement", in: *Hastings Center Report*, 1995, No. **6**, p. 23.

[20] A. Alichniewicz, "Philosophical Thanatology as a Basis for a Medical Art of Dying (Medical Thanatology)", in: M. Galuszka, K. Szewczyk (ed.), *Death Studies of Cultural Bioethics*, Warszawa, 1996.

JÓZEF LIPIEC

EXISTENTIAL VALUES

In modern philosophy, we can see more and more distinctly such approaches to the structure of the world of values, in which purely axiological aspects are preceded and supported by *ontological* statements and discoveries of *philosophical anthropology*. This relates, above all, to the existential structure of a person, in which a specific assembly of his/her "essence" and "existence" takes place. Secondly, there are references to the triple meaning of the concept of "person". A person is not only a psycho-physical *individual* substance, but also a *compound* subject (community, society), or a *typological* one (species, mankind). In each sense a subjective system of reference for values may be recognised.

In this double context, we distinguish – within the axiosphere – *existential* and *essential values*. The issue of values that are called "existential" here is, *nota bene*, frequently neglected in standard axiological dissertations, which usually focus on "essential" values, e.g. moral, aesthetic, sacral or cognitive ones. Existential values refer basically to the real existence of a human being, but surely not only to this. Some values determine the existence of a community, while others determine the existence of all mankind. What is interesting is whether "*common*" values, equally valid for the individual, for social structures and for humankind (at every stage of its development), are possible. This is probably a variation on the question about the relative and absolute nature of values.

WHAT ARE EXISTENTIAL VALUES?

If the human being is a necessary system of reference for all values, then some of them are decisive for his existence. The very fact of the existence of a human subject as the recipient or creator of all values is decisive here. A non-existent person, falling into oblivion, cannot be an active partner for any of the values. S/he can neither commune with them, nor realize their message, nor even create new ones for his/her own or other people's needs. Realization of existential values determines the existence of the subject in the act of *coming into existence, maintaining the process of existence*, as well as all existential states that accompany him/her, including attainment of *existential fullness*. Therefore, positive existential values beneficially influence human existence, while negative ones break and destroy this existence.

A.-T. Tymieniecka and Z. Zalewski (eds.), Analecta Husserliana LXIV, 173–182.

Thus, existential values correspond to Hamlet's question, "To be or not to be?", as well as to less dramatic but more practical issues: what to do and what to use to protect human existence and what to avoid in order not to cease to exist. Essential values refer to the *form* of this being, namely, "specifically human" existence, "characteristic of man", expressing "man's sense and his place in being". Essential values refer to qualities of humanity issuing from the depths of its essence, and, at the same time, to that which distinguishes a person from everything that is not him/her but which may be accepted as a natural or supernatural background (e. g. things, animals, angels and gods). They give an indication of what a person could and ought to be like if s/he aspired to perfection.

Thus, differences are clearly visible. From the essential point of view a person should be, for instance, good, beautiful and wise, and make the world around him/her like this. From the existential point of view, s/he must concentrate on one major problem: *to be*, and s/he must resist the threat of non-being, or of being a handicapped, dependent entity. Furthermore, s/he should *be* him/herself, but, at the same time, promote the *being* of others. To *be* him/herself, but also whole – collective and typical. Thus, the realization of existential values creates the existential basis for other values. The existing human subject constitutes the necessary condition for essential values. In this way the living axiological circle is closed.

To complete the picture of the essential and existential structure of the axiosphere, we must add that there is also a third kind of value, namely "*ornamental*" values. They are connected with pure play, merriment, fashion, conventional habit as well as transitory emotions. Ornamental values sometimes aspire to axiological advancement, and it sometimes happens that they turn into essential ones. They also need, as a prerequisite, a previous realisation of values creating human existence.

Internal and External Determinants

Existential values cover two domains of the real carriers of the axiosphere: (1) *the internal states of a person's structure*; (2) *the system of external conditionings*. Existence is an *autothelic* ultimate value, maybe the supreme one. Simultaneously there are a number of *instrumental* values, closely connected with various determinants of human existence.

Therefore, accepting the existence of an individual as the reference system, the first domain that is positively valuable is that which creates real *life*, and furthermore, life that is efficient and healthy in every respect. Thus, the negative value is *death*, and its particular attributes as well as its signs, like *illness* or *disability*. In the second domain, the positively valuable elements are all states of the surrounding world that are favourable to people and make their existence possible until the totality of natural possibilities is completed. Thus, the domain of external determinants includes: a) all natural conditions like the values of appropriate temperatures, air pressure, oxygen for breathing or life-giving food; b) the values of social-civilizational protection, that is, appropriate institutions and facilities as well as the actions of men equipped with appropriate technological means.

Language and Cognition

There is no complete inventory of basic terms for the existential part of the axiological sphere. Dictionaries are, in this respect, rather poor, reflecting, as it seems, the general axiological situation. Well, existential values, though the most primitive genetically, manifest themselves in cognition relatively slowly, obscurely and selectively. Though many of them have always occurred in the full glamour of ontical value, their discovery as recognised and culturally adopted values has been a long and complicated historical process. Values include usability of oxygen, purity of water and dampness of the ground, but the discovery of these values has been gradual – sometimes they have only been discovered when their absence has led to ecological catastrophe. The history of studies on the causes of illnesses has had a similar significance leading to radical changes in view. Drugs, alcohol and nicotine have always been objective poisons, though only recently has their proper place in the axiosphere been found. There are similar problems in the psychical sphere. Conscious or unconscious stresses are the carriers of constructive and destructive values which are important for man's existence.

The language of axiology does not always keep up with these discoveries. Numerous areas and particular values are expressed in a purely objectivistic or even strictly technical way, apparently devoid of evaluations. Nutrition is a value of agricultural products, just as deadliness is a property of the hydrogen bomb. Yet, what should we call the values of the chemical composition of the atmosphere or of organic food? Should they merely be labelled as "giving a defined amount of energy"? We can also find such an apparent absence of emotions in medical terminology.

The Range of Individual Values

A person's being itself is a fundamental existential value. On an individual level it relates to the existence of (a) a *given*, or (b) *every* individual. *My* existence, as contrasted with another person's, is a special case. In reference to other people's existence I find values close to those that characterise my existence, but also those which are axiologically distant, indifferent or even opposite. This depends on the emotional attitude towards various persons. We treat ourselves and those whom we love in a completely different way from those whom we do not know, and our treatment of the objects of justified hatred is still quite different. We often take a similar attitude (though devoid of personal feelings) towards the existence of persons, due to their essential axiological qualifications. In extreme cases, we simply deny their right to exist – this relates to two main actors in the drama, namely, the *enemy* and the *perpetrator*.

In the general approach, we emphasize the positive value of *all* individual human beings, independent of any of their properties or external references. Thus, every existence is a value, as is that of evil individuals, failures, misfits, including those who oppose life by attempting suicide. Universal, unexceptional treatment of individual existence as such induces serious consequences in the evaluation of the history of the human species and particular societies, as well as a fundamental revision of educational ideals, the state of other institutions. The example of the debate on capital punishment is just the tip of the iceberg. A complete evaluation of military civilisation and culture, which justifies or even glorifies killing, is necessary.

The Value of Life

Thus, the existential value is, first of all, human life understood in various ways: as my life and your life, a particular life or any life, a chosen life or every life, and as individual life as such (i.e. the "idea of human life"). Secondly, existential values include all culminating points of the life process, like the act of creating life (conception and birth), and its sad end. The death of an individual is a total, irreversible destruction of its being. Thus, we should recognize death as an extremely negative existential anti-value. Thirdly, existential values cover the whole area of internal and external conditionings of a given individual or every individual. Therefore, it is not only life as a permanent ultimate produce which is a value, but also the way to this goal and *all life-giving natural and civilizational factors*. Among the latter, a significant role is played by human actions, both single actions and long sequences involving numerous subjects. Existential value is ascribed to surgical operations *hic et nunc*, or to actions such as saving a drowning child, for instance, but also to progress in medical investigations in the general sense. Conversely, anti-value characterises both the crime of the murderer, and the anonymous, blind tendency to destroy fertile grounds or to use nuclear arms.

"Life" is by no means a synonym of "existence", and existential values cannot be simply reduced to life values. The existence of a real person is essentially identical with his (biological) life and independent of the possible opportunities of subjective "extra-life" existence, somehow abstracted from his body. I shall not discuss this question here for obvious reasons: it belongs to the domain of ontological and anthropological studies, and its solution should be sought in the metaphysics of final affairs. Yet, the values determining human existence need not themselves be of "life", if we assume that the universe is not a living organism, but merely an efficient mechanism.

A Healthy and Fit Existence

Existential values turn into qualitatively concretised indices of *effective existence*. Then they no longer relate to a person's existence itself, but to an *internally healthy existence* which is *externally fit*. A life of *existential fullness* proves to be the ideal of existence. At the other extreme, we see various axiologically degenerated and ontologically dependent forms of life persevering, thanks to external feeding. In their extreme forms they are capable only of maintaining some "auxiliary" functions and none of the human, subjective ones.

The autothelic values of health and fitness, close to the Platonic, Olympic ideal of beauty (of bodies and souls) are accompanied by numerous *instrumental existential values*, superimposed on the totality of *means* and *measures* aimed at *maintenance, restoration* or *improvement of a person's health and fitness*. Repositories of existential values conceived in this way include: (a) medicine as a science and art of treatment, together with the related institutions and material background (equipment, drugs, facilities); (b) physical culture (rehabilitation, recreation, sports), and, in a slightly different dimension; (c) science, education and art, shaping the health and mental fitness of an internally harmonised man.

Existential fullness is a distinguished existential value. It can be understood either as (1) a value of a certain state achieved in a definite period of life, usually a short-lived culminating point of the individual's development, or as (2) a set of values proportionally distributed over the whole life, causing it to be successful – when comprehensively utilised. A subjective sign of this fullness is a sense of *happiness*. An objective measure is the level of utilisation of talents given to the person. On both these planes, however, there is sometimes a specific conflict between potential and actions. Life is short and conditions can be unfavourable, so it is impossible to realise one's whole potential. Usually a person must choose: either reaching the top in only some fields in which s/he has potential, or harmoniously fulfilling all his/her abilities to some extent.

Social Values

We shall not solve the controversy of whether the majority of values essential for an individual are a concealed form of *existential values for society or the species*. The important thing is to make individuals adopt behaviours not so much that demonstrate "true humanity", but that are conducive to the unity and safety of the community. Morality and law are intentionally meant to serve the good of the community, and only incidentally do they promote the success and perfection of individuals. In this sense, existential values are relativised to ensure that they serve a given community (or every community).

Perhaps such phenomena as the melting of the polar ice caps and international food shortages will mobilise humankind to "come into being". In such areas, we can see a consensus of values common to individuals and various groups. In other fields, however, we encounter significant discrepancies. The anti-value of individual death is usually treated with indifference by collective entities. What is more, sacrificing one's life in defence of a nation is recognized as a heroic deed deserving admiration, or even an ordinary, indubitable obligation. Restriction of liberty, devotion, sacrifice, suffering, the lack of individuals' self-realization, appear as a logical consequence of the process of fulfillment of the existential mission of a social being. It is, obviously, based on the successful existence of individuals, though it does not support them excessively in definite incarnations. Elements of the whole must be subordinated to the general order. Limited and extremely effective human life (to the point of sacrifice) can be in perfect harmony with other similar lives, composing an efficiently functioning, complex aggregate structure. Care for its safety finally leads to lavishness in the health and lives of its own members and the members of other collective entities. The norm "do not kill" is replaced by the principle "ruthlessly destroy all those who threaten you or may do so". Apart from the positive value of the *common* work of collective subjects, this has probably been the main imperative in the history of mankind, acting for thousands of years like a common law of nature. A physician in a field hospital is a paradoxical symbol of this phenomenon. He saves the life and health of a soldier within the framework of a process aimed – in principle – at killing and hurting. In this situation, medical help is meant to reconstruct men's further capability of killing and being killed, obviously, for the sake of some noble social reasons.

We should state that social existential values regard the existence of real beings of a higher order with a distinctly marked outline of *ontological objectivity*. In their content, they comprise not only many-generational sets of individuals, but also the relations between them as well as common culture and common location. The conditions of existence in, say, Poland, France or Zimbabwe, and all the internal and external threats to their existence, are generally known. Social values are not only a multiplication of individual values, although in some situations, e.g. epidemics, the killing of individuals destroys the basis of existence of a given community. The most important existential problem may prove to be the destruction of culture or the state: the violation of the *ontic identity* of a given social being. In extreme cases, the destruction of a given society may be accompanied by the satisfaction of all the existential needs of its individual members. It is the reverse of the dependencies discussed above. Thus, it is possible to allow the destruction of a given nation without violating the existence of its members.

The Existence of Species

It is most difficult to determine the meaning of existential values with respect to species. This perspective is limited by the horizons of individual consciousness and culture, which always constitute the self-knowledge of a given society. What values refer to the existence of the human being as such? What is an autothelic value here? The ability to persevere in being – despite or within the framework of strong competition, or the ability to develop? For instance, take the tendency to create Platonic Good or Nietzsche's Superman, or Teilhard's Noosphere and unity with the Absolute (or independent self-creation of the absolute). Is it that God is the value of a human being – as some people believe – or is it – as others claim – humankind itself, understood in terms of asymptotically developing "humanity", that is to be the ultimate criterion and axiological goal?

None of these questions can receive an answer that is supported by sufficient proof. It is enough, however, to accept the indubitable occurrence of basic existential values, such as:

(1) Reproduction and hence prolongation of the species' existence.
(2) Genetic improvement as a result of positive and negative selection.
(3) Civilisational development leading to the relative neutralisation of negative factors and mastery over the surrounding forces of nature.
(4) Development of abilities and means warranting external safety.
(5) Attainment of an appropriate quantum of theoretical and practical knowledge, including a surplus that may prove useful in unknown situations in the future.

The set of negative values which includes both inherent, everlasting, external threats and, maybe, irremovable faults in the construction of the human world, is equally clear. Obviously, the most interesting part of the axiosphere is that part which includes, on the one hand, purposeful challenges for the species, and on the other – a false evaluation of means and methods. This relates mostly to *alienated values*. Created by human ambition and hastiness, they are meant to serve people with their own essence, but, instead, they frequently lead people astray, wasting their energy in the pursuit of phantom goals. Then the power of existence weakens, with the consequent threat of ultimate disaster for the species (e.g. nuclear or ecological or some hitherto unknown disaster, not even discernible at the present moment). We have no certain knowledge about the effects that can be brought about by the general "technologification" of the world, or the process of genetic manipulation, or even the modification of the course of evolution through human intervention in various areas (from medicine and technology to culture and education). Existential values relating to the species are obscured, like no others, by a fog of uncertainty and the risk of really vital errors (if we consider the space-time range they cover).

Axiological Compromise and Common Values

Existential values are interrelated in various ways. What is good for the existence of an individual need not be equally good for society. What is favourable for the species does not necessarily agree with the needs of a given individual or community. Aware of these conflicts, we are reluctant to admit our helplessness in solving them. And the incessant search for compromise and common values is always going on.

A common value in its essential properties performs the same effective function for different subjects. Mutual love makes both lovers equally happy. A masterpiece evokes equal admiration in all its perceivers. A mathematical theorem is true in all cultures and for all people. "Commonness" is sometimes *relative* if there exist subjects differently endowed with a given value. The mutual love of a happy couple need not satisfy the rejected admirers. A masterpiece loses its glamour in a new, more refined epoch. Commonness becomes *absolute* when, with each substitution, a given value manifests the same, qualitatively determined constructive or destructive capability.

When a value common to two or more parties is not possible, it is necessary to find a specific, synthetic "third" value – *a compromise.* A compromise brings a solution that is not optimal for each separate party, but is better than a conflict that might lead to mutual destruction. A compromise, a "third value", is an acceptance of a lesser evil and a lesser good. Lesser, but good.

Do existential common values exist? Common to different individuals, to individuals and communities, to different societies, to individuals and the species, to particular societies and mankind in general? Is Life such a value? Maybe it is, but under certain conditions, as mentioned above, e.g. in the context of the individual and the species. It seems that there are only two areas that are difficult to question, covering the *common part of the axiosphere.* We find here (1) a set of objective conditionings that can be described as the *world favourable to humankind,* and (2) the elementary canons of reproduction of human existence included in *love-creationism* (sexual and parental). Common values of humankind perceived in any way, are: sunshine, appropriate air pressure, and harvests. A common value is also the ability to give birth to children and bring them up, and, therefore, to protect and prolong existence – individual, social and that of the species, simultaneously (as a genetic and cultural pool).

Other values occur in the sphere of the incessantly created *axiological compromise*, with the emphasis on turning them into common values. This emphasis is applied, it should be stressed, by essential values, first of all those which present the ideal of humanity in the ontic, genuinely *collective* consensus. This should be aimed at liquidating the inborn and acquired historical conflicts between the individual, the group and the mission of the species. It is obvious that this would bring the end of the Epoch of Humankind (as we understand the word), and Utopia would come into being.

THE POSITION OF ESSENTIAL AND EXISTENTIAL VALUES IN THE HIERARCHY OF VALUES

We should also mention the competition between essential and existential values for primacy in the hierarchy. Philosophy started by ascribing this primacy to selected essential values. Existential values tended to be put aside, demanding, from time to time, serious consideration and theoretical discovery. This happened from many reasons.

What reasons? On the one hand, an awareness of the serious threat to the world and people was born. On the other, was the gradually emerging ability of people to exert real influence over their fate, that is, its condition and duration. Without diminishing the significance of the mood of increasing fear, we must stress the importance of the factor of *pragmatic optimism*, and of the *subjective responsibility* connected with the discovery of a *chance for success*.

As long as a person has no influence on the course of his/her being, s/he must humbly accept all existential transcendent verdicts. At a proper moment, not chosen by him/herself, the person should die with dignity, honour and for a noble cause, before God, the nation and nature. Existence gives room to the essence, for it is the only rational solution. However, when we can see that it is possible to defend oneself from a premature, nonsensical falling into non-existence, when it is possible to avoid the inevitability of suffering, when we can saturate our lives with various goods and a sense of happiness, a radical change in propositions and positions could take place. Existential values would cease to be a shameful addition that should not be mentioned in sublime considerations, and gain a significant place in the axiosphere. Depending on the cognitive point of view, either as the basis of the whole conventional pyramid, or as its top, they would show their primacy in each case.

Jagellonian University

MAURICE A. M. DE WACHTER

IS PERSONALISM AN ADEQUATE MORAL SYSTEM FOR BIOETHICS?

The Test Case of Artificial Insemination

1. IS PERSONALISM USEFUL IN MEDICAL ETHICS?

Many disciplines, besides ethics, have considered the rapport between bioethics and personhood. Historians and sociologists, for instance, have concluded that bioethics and the concept of personhood need each other.[1] As a bioethicist I would like to ask an additional question – to what extent do personalist morals really meet the needs of bioethics?

Personalism in ethics means any ethical system that is based on a person's values. Personalism is also a generic name for several, sometimes quite different, views of the human being that have originated in philosophical, theological and secular schools of Western thought. There are, nevertheless, two major definitions of personhood in Western history: one is Christian in origin, and the other secular, more particularly Kantian, in origin. The Christian definition sees the person as a religious being. This definition is mostly linked to the original view of Boethius, a sixth-century philosopher who saw the person as an individual substantial being of a rational nature, that is, an individual gifted with reason. Immanuel Kant saw the person as mere will. Both definitions are at work in today's schools of personalism. It seems, though, that the Kantian definition of a person as "an ethical being with freedom and responsibility" dominates Western ethics nowadays.

Within today's Catholic moral theology there are at least three types of personalism.[2] Common to all three is their European origin as well as the acknowledgement of the basic and central notion of the person in ethics. In all three types the person is a union of bodily and spiritual components. Differences appear in method, especially in the way in which the concept of a person functions as a norm. In the *first* version, for instance, which is a type of personalist anthropology, the description of moral reasoning has not changed. It remains deductive, confirming and further legitimising previous norms. The *second* version essentially adheres to "proportionalism" – a method which includes intention, act, and consequences for the person, and weighs personal values gained against values lost. This calculation makes it possible for moralists to revise previously held positions. The *third* version of personalism builds upon personal moral awareness through intention and

A.-T. Tymieniecka and Z. Zalewski (eds.), Analecta Husserliana LXIV, 183–192.

creative conscience. With its interest in various zones of personhood, going from the core of fundamental options to peripheral and superficial moral commitment, this third version also seeks to embrace human reality at its fullest.

In my paper I will talk only about one well-known type of personalist morals – Leuven personalist morals. So-called Leuven personalism is a form of personalist morals being developed in the Faculties of Philosophy and Theology at the Catholic University of Leuven since the 1930s. It has two of the above mentioned bases and characteristics: first, it feeds on both religious and secular sources; second, in line with the second version, it uses proportionalism as a system of weighing moral values. Being a mix of religion and secularism, Leuven personalism shows religion and secularism to be complementary, but sometimes it also reveals their rivalry. The dominance of faith in theological sciences and of reason in philosophy and human sciences, lead respectively to either a religious or a secular concept of personhood. Often their rivalry remains undecided and gives birth to a hybrid which repeats all the well-known tensions between faith and reason. A tension appears between convictions coming from belief and the necessary rational clarification of those convictions. From experience we have learned that the process of clarification readily shows that only the strength of an argument, and not authority, determines the worth of the moral conclusion, no more, no less. The same process of necessary clarification possibly creates epistemological heuristic problems that may weaken personalist morals.

On the basis of the remarks offered in this introduction, I should like to raise two questions regarding Leuven personalist morals. The *first* question is as follows: can a system that displays a chronic tension between religion and secularity ever be useful for medical ethics in a pluralistic society? The *second* question, illustrated by a test-case in the area of artificial reproduction, concerns the scope, the width and the applicability of this version of personalism in today's pluralistic society.

2. LEUVEN PERSONALISM: THE HUMAN BEING AS THE MORAL NORM

Of the three versions of personalism that seem to be at work in Catholic moral theology and that were mentioned earlier, we now focus on the second. Here, personalism functions as a system based on the dignity of the person, and the norm is to be found in the human being. The morally good act corresponds objectively to "the human being considered in the light of reason and of faith."[3] In 1979, the main representative of Leuven personalism, Louis

Janssens, published an article on artificial insemination. In it he stressed that ethics increasingly needs insights coming from human sciences such as psychology, in order to better understand human acts and their consequences. Janssens was aware of the weight given to scientific data as well as of the differences in the foundations underlying the systems. Janssens' preference is for "an ethic of responsibility based upon the human being."[4]

The Vatican Council, on the subject of human procreation, states that "the moral character of human action ... should be defined by objective standards which are based upon the nature (being, own character) of the person and of the acts of the person."[5] Janssens then goes on to say that here a general principle of morality is being stated. This means that its application is not limited to sexuality, but applies to all areas of human activity. Janssens also insists on the mutual interrelatedness of all dimensions of the human being. These dimensions can be described as follows:

1) The human being is a subject, not an object.
2) S/he is an embodied subject.
3) S/he is also part of the material world.
4) S/he is essentially oriented towards other persons.
5) S/he needs to live in communities, social structures and institutions.
6) S/he is called to know and to love God.
7) S/he is a historical and growing reality.
8) While being fundamentally equal, human beings are at the same time unique and original.

Janssens considers it an absolute prerequisite, for his personalism to work, that all eight dimensions be taken together. Only together do they represent "the only whole that every person is."[6]

3. ARTIFICIAL INSEMINATION: A TEST CASE, 1979

Personalism and Sexual Morality

Leuven moralists had two reasons for adhering to personalism: respect for the human experience of Catholic married couples, and criticism of the physicalist thought pattern. They were convinced that sexual morality could not be meant to make people's lives miserable. Taking human experience seriously and acknowledging the validity of empirical knowledge inevitably brought them to revise previous moral positions. This revisionist project applies particularly to the traditional "morality of acts" where the norm is sought in conformity with biological norms. This "morality of acts considers sterility to be the result of a (God given) nature" which has willed that couple X or Y shall not have children.[7] A shocking example of biologically determined moral judgements was the pronouncement of Pope Leo XII in 1829 regarding vaccination against smallpox. "Those who lend themselves to be vaccinated," so the text goes, "stop being God's children. Smallpox is a judgement of God, to vaccinate is to defy heaven."[8] Another example of biologism as a moral norm is the 1946 article on artificial insemination by Hurth. He says: "La nature a imprimé et dans la structure du *semen* et dans celle de ce mécanisme (le déclenchement physique et psychologique du mécanisme sexuel) une finalité exclusive, comme aussi une manière exclusive de la faire atteindre pour le bien de l'espèce."[9] Instead of relying on biological criteria, personalist thinkers look at the mutual surrender of marital partners and at procreation in line with human dignity. Both these norms open the possibility of at least considering artificial insemination, be it homologous or heterologous.[10]

Leuven A.I.D. Practice in the Sixties and Seventies

In the University Clinic of the University of Leuven, the first Artificial Insemination by Donor (A.I.D.) was performed in August of 1971. By 1983, between 60 and 100 couples were inseminated every month.[11] Particular to the Leuven practice was the presence of a (counselling) team. Multidisciplinary in its constitution, the team included several gynaecologists and obstetricians, an andrologist, a urologist, a psychiatrist and an ethicist. The aim of the teamwork was to create and maintain optimal conditions for quality treatment. One such essential condition was that the couple would be able to distinguish between biological and human parenthood. Indeed, not every biological generation produces of itself a human level of parenting, and vice versa. After several years of experience, in 1976 the team stated its mission as follows: "In well-defined cases of undesired childlessness where, after serious investigation, after serene deliberation and a mature choice, and where the possibility exists of offering a higher level of quality of life as well as of fulfilment of life, we deem it ethically acceptable to apply the treatment of A.I.D. with utter care and sense of responsibility."[12]

Personalism and A.I.(D.) in the Seventies

As mentioned before, Louis Janssens, one of the main representative personalist thinkers of the Leuven school, brought the theory of personalism to bear upon the practice of artificial insemination.

Leuven personalism justified homologous artificial insemination in vivo on the grounds of such personalistic values as the well-being of the husband and wife, the well-being of the child, the well-being of society.

Regarding heterologous insemination, Janssens saw "the moral question (to be) about valid reasons (ratio proportionata) that would justify fertility treatment and whether, according to rules of priority, what will be realised outweighs the negative aspects." Despite initial emotional reactions, the moral evaluation must focus on whether or not we have here a practice worthy of human persons. Married couples, according to all evidence, experience this procedure not as a threat, but rather as enrichment. Couples who donate sperm see this as an act of kindness to others. In 1960 up to 93% of all gynaecologists opposed A.I.D., whereas in 1970 this percentage had come down to 42%. "This convergence of numerous positive experiences ought to, at least, incite moralists not to condemn radically the practice of heterologous artificial insemination."[13] The final words of Janssens in a 1979 article were, "no radical condemnation, but a critical selection of both candidates for treatment and donors."[14]

4. PERSONALISM AND BIOETHICS

How useful then is personalism in today's bioethics? We shall try to answer this question in two steps. First, we shall consider the strong points of personalism, as well as its weaknesses as a moral system. Next, we shall try to verify to what extent personalism is consistent in its application of norms, in particular with regard to the case of artificial insemination.

The Strengths and Weaknesses of Personalist Morals

Personalism generally tends to abandon moral systems built on acts and their meaning, especially in the restricted biological finality of these acts. Personalism, even when based on religious and theological grounds, affirms the need to trust human experience as a source of morality and of moral norms. "The meaning of experience and the absolutely essential demand for interpretation of experience inevitably led to more metaphysical and sociological considerations."[15] Methodologically, then, induction has an important place. Human sciences, psychology, sociology, cultural anthropology, political sciences, must be part of the moral debate and part of the moral system. This conclusion is one of the major contributions made by personalism to the new covenant between revelation and reason.

However, personalism is not without its weak points. As a normative system, personalism has been found to suffer from the same weaknesses as the system it is trying to correct, i.e. the system of physicalist and biological morals.[16]

There are three weaknesses that have been identified: the first is its inadequacy in representing the whole person; the second is the lack of coherence with the totality of the person; the third is the limited moral validity of the acts of the person.

First, the insufficiency of a morality of acts lies in its limitation to the biological dimension of these acts only. According to personalist morals, this gap should be filled with the remaining aspects of the human being. However, as yet there is no univocal definition of the notion of "the person integrally and adequately considered." There remains doubt about the number of other personalist aspects that ought to be taken into account before one can talk about reaching personalist standards. Moreover, we still do not have one single all-embracing theory of the human being. No systematic description of the moral status of each of the so-called personalist dimensions is available. It would be important, for instance, to know whether we are talking about phenomenological descriptions or about ontological qualities. Equally important would be a scale indicating the hierarchical order of those personalist aspects.

Second, the same criticism applies to the coherence of various personalist dimensions. Their normative weight, as well as the criteria whereby it is possible to determine what is truly promoting the human person, remain unclear. For instance, human sterility/fertility is certainly an important dimension leading to frustration or fulfilment in the field of procreation and parenthood. Surely, we should try to prevent or remedy such frustrations and offer fulfilment. But at what price and at what human cost? Here, personalism rightly calls on the doctrine of "proportionalism," but without providing "practical rules of thumb" that would be of help to those working in the field.

Third, like any other normative system, personalism ought to clarify what it means when the human being is made into the criterion of right and wrong. For instance, does the norm of "the person and all his relationships" simply do away with the past biological criteria of human acts, or should it still take them into account? And if so, to what extent? These are questions that need further exploration and satisfactory answers before personalism is to be fully recognized as a moral system.

The Test: Ambivalence Regarding A.I.D., 1990

Coherence and consistency are essential qualities of any moral system. If one author answers in the positive where another answers in the negative, we must scrutinise the reasons for such a dichotomy. The North American moralist Richard A. McCormick has openly acknowledged that Louis Janssens is the person who has best explained what it means to use "the human person integrally and adequately considered" as a criterion.[17] Nevertheless, while accepting Janssens' list of personalist dimensions, McCormick states that the personalist norm becomes particularly delicate and controversial in the area of sexual ethics. In this area, more than in any other perhaps, humans are always "liable to self-interested judgements and insensitivity to the call of basic values."[18]

The difference in moral position between Janssens and McCormick appears in a policy document of the Ethics Committee of the American Fertility Society, published in 1990, called "Ethical considerations of the new reproductive technologies." McCormick, a member of the committee, visibly included personalist principles in this document. He also pointed out that the personalist system may have its weaknesses, in particular a "moral ambiguity and pluralism (which) calls for openness, caution and a willingness to revise evaluations."[19]

The Ethics Committee is aware of the fact that the use of donated sperm still causes serious controversy. Nevertheless, the practice of A.I.D. is deemed acceptable. However, Appendix A – "Dissent on the use of third parties" – seems to have been written by McCormick. It states that "the use of third parties – whether by donor sperm, donor ovum, or surrogate womb – is ethically inappropriate." This statement is supported by five objections to the use of third parties:

First, the use of third parties seems to violate the marriage covenant. It fundamentally severs procreation from the marital union.

Second, it brings into the world a child with no bond of origin to one or both partners and therefore blurs the child's genealogy and potentially compromises the child's self-identity.

Third, if the moral right to use third parties were conceded, wives might easily conclude that it is preferable to be inseminated the natural way. Thus, the incidence of adultery might increase.

Fourth, the "stud-farm mentality" is accompanied by a subtle but unmistakable move towards eugenics.

Fifth, it tends to absolutise sterility as a disvalue and child-bearing and rearing as a value, thus distorting some basic human values: life, marriage, and the family.

The author concludes: "Taken cumulatively, such considerations suggest that the use of third parties to overcome sterility is *not for the good of persons integrally and adequately considered.*" (Emphasis added.) "It involves risks to basic dimensions of our development. Such risks to basic values outweigh, in a prudent calculation, individual procreative desires or needs. To summarise: when the calculation involves individual benefit versus institutional risk of harm, the latter should take precedence."[20]

All the personalist elements are at work in this conclusion: the value of the person integrally and adequately considered, as well as the calculation, that is, the method of proportionality. Yet the outcome is opposite to what Janssens accepted about artificial insemination using donated sperm. The one point that does seem to cause a discrepancy is tension between the individual and the collection of persons in society.

Perhaps one should say that both authors agree to disagree. McCormick, in the introduction to the Ethics Committee's considerations, already said the following:

> Some actions remain ambiguous because they involve both beneficial and detrimental aspects, because their impact on persons is unknown, or because they are variously evaluated. For example, some persons will see the benefit to a single person or couple as dominant; others will view the potential risks to society as a whole as taking precedence over individual benefits.[21]

And, indeed, Janssens also already said in 1979 that the diversity of opinion may "depend on the degree to which people take into account the facts of science, but also the principles of ethics, and the way they approach problems."[22]

5. CONCLUSION

At the outset of this paper we raised two questions about the usefulness of personalist morals in medical ethics. Our first question was: "Is this personalism, given its internal tension between religion and secularism, useful outside the circle of its own adherents?" The other question was about the scope of personalist morals as tools for bioethics. Here, artificial insemination was the test case.

The first question is answered by looking at the diversity of opinions expressed by two important authors. How useful can a moral system be when it allows "yes" and "no" answers to one and the same moral issue? The question of whether personalist morals can be generalised has been answered by Janssens, who claims that the personalist standard is objective in the context of human sexuality, and can even be seen as "a general principle which is valid for all domains of the human activity."[23] Truly, the use of experiential knowledge as a source of personalist norms makes the system quite attractive even for those outside the community of believers. And, last but not least, the perception of the human person as a subject-in-the-world as well as the view of the person as "originality" make personalist morals attractive even for existentialists.

With regard to the second question as to whether personalism is complementary within a pluralistic debate, serious warnings must be given. A method which maintains that human reason needs revelation makes personalism inherently suspect to anyone who stays outside the realm of faith. This is a threshold which few, if any, non-believers are able to cross.

The overall conclusion, then, may be that Leuven personalist morals have undoubtedly had a profound impact upon medical ethics within medical ethics circles. The very fact that personalist morals have been introduced in Church documents at the highest level shows that they have tended to be generally accepted. Here lies the strength of Leuven personalism. Less convincing impressions are derived from inconsistent conclusions regarding practical issues of medical ethics, even within medical ethics circles. Therefore, the probability that Leuven personalism will develop into a universally useful system of bioethics is obviously low. Its merits remain restricted to those of the inner circle, and perhaps to some others who believe that the demands of reason sufficiently warrant a dialogue within personalism.

Waterloo, Belgium

NOTES

1 R. Fox, "Personhood, Medicine, and American Society," *Millbank Memorial Fund Quarterly/Health and Society*, **61** (1983), No. 1, pp. 127–147; W. de Craemer, "A Cross-Cultural Perspective on Personhood," *Millbank Memorial Fund Quarterly/Health and Society*, **61** (1983), No. 1, pp. 19–34.

2 B. V. Johnstone, "The Revisionist Project in Roman Catholic Moral Theology," *Studies in Christian Ethics*, **5** (1992), p. 29.

3 L. Janssens, "Personalist Morals," *Louvain Studies*, Fall 1970, pp. 5–16.

4 L. Janssens, "Kunstmatige inseminatie. Ethische beschouwingen," *Verpleegkundigen en gemeenschapszorg*, **35** (1979), p. 220; the same in English: "Artificial Insemination: Ethical Considerations," *Louvain Studies*, **8** (1980), pp. 3–29.

5 L. Janssens, op. cit., p. 221. Quotation from Pastoral Constitution, *Guardian et spes*, 1965, No. 51.

6 L. Janssens, op. cit., p. 221.

[7] M. A. M. de Wachter, I. Brosens, P. Nijs, A. van Assche, R. Vereecken (eds.), *Menselijke vruchtbaarheid en geboortenplanning. Het paar en zijn begeleidend team*, Brussels: Elsevier Sequoia, 1976, p. 119.
[8] L. Janssens, op. cit., p. 227.
[9] F. Hurth, "La Fécondation artificielle. Sa valeur morale et juridique," *Nouvelle Revue Théologique*, **69** (1946), pp. 402–416.
[10] M. A. M. de Wachter, "Une Procréation à la mesure de l'homme," in: *Echanges*, **115** (1974), pp. 29–31.
[11] O. Steeno, "Elf jaar praktijkervaring met 509 zwangerschappen door donorinseminatie (K. i. D.)," *Actualiteiten: relatie en seksualiteit*, 6 (1983), pp. 8–12; I. Brosens, O. Steeno, P. Nijs et al., *Donor-inseminatie. Leuvense Cahiers voor Seksuologie*, Leuven: Acco, 1977. See also: K. Demyttenaere, P. Nijs, O. Steeno, Ph. Konincks, and G. Evers-Kiebooms, "Anxiety and Conception Rates in Donor Insemination," *Journal of Psychosomatic Obstetrics and Gynaecology*, **8** (1988), pp. 175–181.
[12] M. A. M. de Wachter et al., 1976, op. cit., p. 114.
[13] L. Janssens, op. cit., p. 241, refers to L. I. Swabb, "Kunstmatige inseminatie": *Nederlands Tijdschrift voor Geneeskunde*, **116** (1972), 2381 vv; and L. I. Swaab, "Resultaten en ervaringen met donorinseminatie," *Nederlands Tijdschrift voor Geneeskunde*, **118** (1974), pp. 493–498.
[14] L. Janssens, op. cit., p. 241.
[15] J. A. Selling (ed.), *Personalist Morals. Essays in Honor of Professor Louis Janssens*, Leuven University Press, 1988, 4.
[16] B. V. Johnstone, "From Physicalism to Personalism," *Studia Moralia*, **33** (1992), pp. 71–96, especially 91–95.
[17] R. A. McCormick, *The Critical Calling*, Washington: Georgetown University Press, 1989, p. 14.
[18] Ibid., p. 15.
[19] American Fertility Society, "Ethical Considerations of the New Reproductive Technologies," in: *Fertility and Sterility* (Supplement 2), **53** (1990), No. 6 (June 1990), pp. 1–109.
[20] Ibid., p. 57.
[21] American Fertility Society, op. cit., p. 1.
[22] L. Janssens, op. cit., p. 220.
[23] L. Janssens, op. cit., p. 221.

PART III

ETHICS OF MEDICINE

H. TRISTRAM ENGELHARDT, JR.

FROM ABORTION TO EUTHANASIA: RETHINKING THE VALUE OF LIFE AT THE END OF THE CHRISTIAN AGE

I. AN INTRODUCTION: WHY IT IS SO HARD TO DISCOVER THE VALUE OF HUMAN LIFE

Many have observed that we are passing through a period as significant as the Reformation and the Renaissance. The very frame of reference by which we understand the meaning of reproduction, birth, suffering, and death is being recast. That we are in a period of significant change seems manifest. Almost overnight, an entire ideological framework with its metaphysical and historical presuppositions collapsed in Eastern Europe. With that collapse, or at least with the substantial reorientation and rearticulation of its claims, a set of assumptions regarding the place of the state, the role of medicine, and the meaning of individual life was recast as well. An analogous transformation has been occurring in Western Europe and North America with regard to Christianity, rendering it ever more a historical root of the culture than a living religious force. In the late 1960's, probably associated with the rupture of pieties and traditions encouraged by Vatican II, the culture of Roman Catholics, and because of their influence even that of the Protestants, became ever more disconnected from the Christianity of the first millennium. Moreover, Christianity became increasingly less influential in personal life and public policy. In a world after communism and Christianity, in a world in which these are no longer dominant cultural forces, one faces the challenge of giving direction to medicine's interventions in human reproduction, birth, suffering, and death. These changes are of particular significance for Poland with its historical connections to both communism and Catholicism.

In medicine, an index of these moral and cultural changes with respect to Christianity is found in the widespread acceptance of medical practices traditionally condemned by Christianity. The increasing acceptability and use of abortion, third-party-assisted reproduction using donor gametes, as well as physician-assisted suicide or euthanasia mark a watershed shift in the evaluation of human life. This is not to suggest the absence of abortion and physician-assisted suicide in even the most pious periods of European civilization. It is rather that increasingly these terminations of human life are regarded as taken-for-granted, appropriate medical procedures. All of this

A.-T. Tymieniecka and Z. Zalewski (eds.), Analecta Husserliana LXIV, 195–207.

has occurred, though abortion, physician-assisted suicide, and euthanasia are at variance with ancient and substantial Christian understandings.[1]

As I will argue in this essay, this state of affairs should not be regarded in merely historical or cultural terms. It has roots in our epistemological condition. We confront the inability of secular morality in general, and of secular bioethics in particular, to justify a particular canonical content-full understanding of the value of life. Outside of a particular moral narrative or particular moral content, persons meet as naked sources of authority such that proscriptions against voluntary active euthanasia or physician-assisted suicide cannot be either justified or indeed understood as compelling. Because of the centrality of persons as sources of moral authority, a distinction arises between human personal and human biological life, which undercuts a secular attempt to justify traditional Christian proscriptions of abortion.

In all of this, I am not an advocate of abortion, physician-assisted suicide, or euthanasia. Quite to the contrary. Rather, the substance of my argument is that the evil of these interventions cannot be understood in general secular moral terms. Indeed, when we turn to the question of the value of human life, we find ourselves asking a question that cannot be answered in general secular terms. To the question, "What is the value of human life, such that this value should be recognized by all rational persons," it is not in principle possible to discover an answer that is both content-full and generally binding. The more an answer is general, the more it steps back from a particular content and perspective. The more an answer has content, the more it involves taking a particular perspective. So, too, all content-full answers regarding the value of human life presuppose a particular perspective, orientation, or narrative. The aspiration to a universal answer, to a general moral rationality binding on persons as such, involves attempting to respond to a question posed as if from nowhere by way of an answer that presupposes a very particular somewhere. This point has been developed at some length in a forthcoming book by Kevin W. Wildes (Wildes). At stake is a category mistake: one asks a question from nowhere in the hope of transcending particular moral contexts and narratives, while at the same time demanding an answer as from a somewhere, that is, from a perspective replete with content-full meaning and direction. One wants moral direction universal in governance, but particular in content.

For example, this shift between the character of the question and the answer with respect to generality and content occurs in the history of terms such as "sanctity of life," a term that at best extends only from the beginning of this century. Or rather, when one does find earlier usages, as with Jeremy

Taylor's *The Great Exemplar of Sanctity and a Holy Life*, published in 1667, sanctity identifies holiness as that in Jesus Christ (Khushf, 1996, pp. 293–310). Appeals to the sanctity of life appear to be attempts to state in general secular terms a claim that can only be understood in very particular religious terms. That is, they function as an attempt to make a claim that should be recognized in general moral terms, but that still has a very particular content. When we seek to articulate a universal value that we should recognize in human life, especially in the conduct of medicine with regard to human life, we find that this cannot be done outside of a particular history and a particular moral account that can justify this particular content. In asking after the value of human life, we hunger after an important answer, which cannot be articulated in general secular terms.

The general problem has special relevance for Polish reflections on medicine. Bioethics in Poland finds itself within the span of only a few years experiencing the full force of the post-Enlightenment fragmentation of a post-Christian, post-communist world, while still keeping intact much of the expectations of the Christian Age. For those in Poland whose religious beliefs are still intact, the character of this problem may appear out of place, if not exotic. Yet as Poland and Polish medicine enter the surrounding post-modern secular culture, the problem of giving a value to human life will be a challenge for Polish culture, Polish bioethics, and Polish health care policy. What now exists as a cultural challenge elsewhere appears like storm clouds on Poland's horizon.

II. REEXAMINING THE VALUE OF HUMAN LIFE AT THE END OF THE MILLENNIUM

An examination of the value of human life for medicine in Poland and elsewhere must be undertaken against the backdrop of the catastrophes of this century. At no time in human history have more people been killed over a shorter period of time. Poland has unfortunately experienced much of the worst of this carnage on its own soil. The killing was justified in terms of a future which revalued current lives in terms of human goals. Millions were slaughtered, millions of whom were workers and peasants, in the name of justice or equity, or on behalf of a better and "more truly human" future. From the French Revolution to the October Revolution, blood has been invoked to purify the present from a Christian past in order to take hold of a good-natured future. As Merleau-Ponty (1908–1961) declared, "Revolutionary justice appeals to the future as its standard" (Merleau-Ponty, 1947, p. 30).

It is remarkable with what naiveté we approached the catastrophes of this century. Despite the French Revolution, the nineteenth century was marked by a confidence quite contrary to the pessimism of Dostoevsky regarding the results likely to follow from a loss of religious orientation and from a confidence in human power. There beckoned, instead, the prospect of a *religion de l'humanité*, to take a phrase from August Comte (1798–1857). There was a widespread expectation of a union of religion and science that would secure perpetual peace and progress. There was a faith in humanity, which often took on many of the marks of religion. As one American pundit opined in an address on May 30, 1867, shortly after the American War Between the States,

> Brothers and sisters, we want to work for humanity. We have a new gospel to proclaim – the gospel of religion and science, two in one – the gospel of faith in man carried out to its extremest consequences. ... We have a new gospel of good news, a radical gospel, the gospel of the "enthusiasm of humanity." God grant us ... a new Pentecostal outpouring of courage and fidelity to truth (Abbot, 1867, pp. 37–40).

There was a robust expectation that, in a world guided by science and freed from superstition, peace and prosperity were inevitable.

The twentieth century began with these expansive hopes. In its first decade one could still make secularly pious and confident expressions regarding the future, such as the following:

> But the world is growing better. And in the Future – in the long, long ages to come – IT WILL BE REDEEMED! The same spirit of sympathy and fraternity that broke the black man's manacles and is today melting the white woman's chains will tomorrow emancipate the working man and the ox; and, as the ages bloom and the great wheels of the centuries grind on, the same spirit shall banish Selfishness from the earth, and convert the planet finally into one unbroken and unparalleled spectacle of PEACE, JUSTICE, and SOLIDARITY (Moore, 1906, pp. 328–9).

In Poland, much of this was anticipated in philosophical and chiliastic terms. The progressivist aspirations of peace, liberation, and tranquil development went aground on the battlefields of the First World War.

The shadow of the October Revolution and the development of fascism brought enslavement and terror rather than freedom and peace. Still, the hope for progress remained firm. One finds Merleau-Ponty defending elements of this carnage:

> For it is certain that neither Bukharin nor Trotsky nor Stalin regarded Terror as intrinsically valuable. Each one imagined he was using it to realize a genuinely human history which had not yet started but which provides the justification for revolutionary violence. In other words, as Marxists, all three confess that there is a meaning to such violence – that it is possible to

understand it, to read into it a rational development and to draw from it a humane future (Merleau-Ponty, 1969, p. 97).

Martin Heidegger (1889–1976) makes an analogous appeal in a lecture during the summer semester of 1941.

'Workers' and 'soldiers' open the gates to reality. They bring about the refashioning of the fundamental structures of human creativity that were previously called 'culture' (Heidegger, 1981, vol. 51, pp. 57–58).

Knowing truly the nature of justice or of proper human order, one could with bloody force impose a new order. Knowing how the future should be, one could coercively address the present.

The difficulty has been that each new history and account is but one among others. Secularity has provided as many accounts of the value of life, not to mention justice, fairness, and equality, as there are major religions. However, unlike the Judeo-Christian religions that found the genesis, justification, and motivation united in God, secular morality could neither establish a canonical content, nor easily unite motivation and justification. Secular morality faced a significant challenge: how can one show that that which one ought to do is that which will achieve the good of the individuals so obliged? How can one show that happiness will come to those who act so as to be worthy of happiness? Indeed, how can one show that it is rational to act so as to be worthy of happiness in circumstances when this will likely bring the opposite of happiness? If secular morality could deliver content-full moral guidance in the sense of justifying a content-full morality, could it also show the rationality of resisting temptations to bad faith, immoral compromise, and betrayal? Could it establish the rationality of moral motivation in the face of significant personal costs to the actor?

Unfortunately, the history of this century shows how hard it is for people to resist immorality when such compromise appears very much in their own best interests. Indeed, when the costs may be as high as life itself, how can one justify a motivation to secular martyrdom? It is because of such challenges, among others, that Immanuel Kant finds himself constrained in the *Second Critique* to act as if there were a God and as if man were immortal.[2] As Kant argues, "The doctrine of Christianity, even when not regarded as a religious doctrine, gives at this point a concept of the highest good (the Kingdom of God) which is alone sufficient to the strictest demand of practical reason."[3] Kant recognizes that at least an "as if" deep integration of morality and nature must be achieved if the secular moral project is to be both universal and rationally compelling. Kant thus attempts to give an integrated account of the right and the good, so as to reestablish the moral enterprise that had fragmented with post-Reformation Christianity into competing moral narratives and communities.

The modern philosophical project, of which Kant was in great measure the exemplary defender, sought by rational argument to provide a morality that could substitute for the fragmented and weakened Western Christian moral vision. If sound rational argument could have delivered a content-full canonical moral account of the right and of the good, one could then have: (1) dismissed all who did not accede to that morality as irrational, (2) constrained them to submit to that rationality with an authority derived from reason, while (3) understanding such coercion as not alien to the individuals subjected to it, but rather restorative of the true rational and autonomous nature of those coerced. In addition, (4) one could have disclosed a rational communality of persons, a kingdom of ends transcending religious and ethnic barriers, uniting all in a content-full rational and moral vision. In addition, (5) one might even have been able to establish a practical reintegration of morality and nature.

In this fashion, philosophy could have replaced with reason the lost focal point of being, morality, and truth once found in God. The rational point of view would then be the moral point of view. Moreover, this point of view would disclose the content that bound together the community of persons as rational agents. A substantive morality could then be discovered as an ingredient in persons as rational agents and that substantive morality could guide human action as well as public policy. Moreover, it could provide substantive understandings of the importance of human life and, as with Kant, lead to arguments against suicide, not to mention physician-assisted suicide. Here it is important to recall that Kant's appeal to autonomy is not an invocation of mere consent or permission as a source of moral authority. For Kant, autonomy is the realization of the rational will, which involves duties to self and brings with it moral content. The content, it turns out, is extensive. As one reads through Kant's *Metaphysik der Sitten*, one finds him reconstructing much of Western Christian morality, including a condemnation of masturbation, a practice he considers on a par with suicide.[4]

The attempt by reason to establish a content-full moral account has failed in fact and in principle. We are left, with respect to secular accounts of justice, fairness, and moral probity, in the same situation as the West was with regard to religion after the Reformation. In addition, the religions that framed the background moral narratives of Western Europe are becoming ever more detached from the lives of Europeans and from the traditional roots of those religions. It is in reaction to this circumstance that Pope John Paul II has raised a lament, decrying both the de-Christianization of culture and the lack of regard for human life.[5] Against this fragmentation of tradition and the failure of reason, bioethics faces the challenge of giving an account of the value of human life. Bioethics, including bioethics in Poland, confronts not only the atrocities of this century, but post-Christian accommodations to third-party-assisted reproduction, abortion, physician-assisted suicide, and euthanasia. In the absence of an encompassing and canonical narrative that would give an account of the value of human life, one finds only persons with different histories torn between competing moral accounts, sharing no common moral narrative. Though Poland may think it still possesses a compelling content-full morality, it cannot secure it in general secular terms.

III. PERSONS, REASONS, AND HUMAN BIOLOGICAL LIFE

If a secular morality is to bind persons across particular moral communities, religions, narratives, and traditions, then in principle it cannot by discursive argument supply moral content or justify a particular a priori value for human life as such. This is not merely a cultural fact or a sociological condition descriptive of post-modernity. It rather reflects the inability of establishing a particular canonical moral vision without either begging the question or engaging in an infinite regress. This is the case because, other than ruling out certain accounts as logically incoherent, in order to choose among alternative content-full accounts of the good and the right, one must already possess a guiding moral sense. To choose the correct moral sense, one must already in the background possess a guiding moral vision, in order to affirm the correct moral sense, thin theory of the good, set of canonical moral intuitions, understanding of moral rationality, etc.

The general character of this dilemma can be illustrated by reviewing a number of the approaches through which solutions are often sought for moral controversies in bioethics and health care policy. If one attempts to resolve a controversy by an appeal to consequences so as to judge which approach will achieve the greatest balance of benefits over harms, one must already know how to compare different benefits and harms, how appropriately to correct preferences, as well as how to discount for satisfaction over time. In short, one must already possess a moral vision. If one appeals to a disinterested observer, no choice will be made if the observer is truly disinterested. A choice will be produced only if the observer is provided with a particular set of moral intuitions, a particular moral sense, a particular thin theory of the good or the right. Any appeal to a hypothetical chooser or set of hypothetical contractors in order to produce a moral vision must already presuppose a background moral vision. The same is the case with respect to appeals to moral rationality. Moral rationality must already incorporate a particular content in order to weight certain value choices over others. Worse yet will be appeals to intuitions. Any set of intuitions, including intuitions about how to regard intuitions, can always be met by contrary intuitions, as well as higher-order intuitions about how one should regard those intuitions.

Outside of a particular moral vision, understanding, or narrative, one lacks a common framework in terms of which moral controversies can come to a conclusion. This is a difficulty with respect to public policy in general, and with respect to bioethics and health care policy in particular. With regard to placing a value on life, one must determine how to compare the value and/or standing of vegetable life, animal life, and human life. With regard to human life, one must compare human life which is part of the life of persons as moral agents with human life which is not. In particular, one must decide how to assess the value or standing of zygotes, embryos, fetuses, infants, humans in persistent vegetative states, etc. How one determines the rank and standing of such instances of life will determine both the obligations one has to forbear from intrusions on that life, as well as to provide support and aid. One will need to have an account of the standing of different instances of human biological life in order to determine whether it is morally inappropriate, for example, to use fetal tissues or to perform abortions. One will need to know the value and standing of individuals in persistent vegetative states in order to determine what, if any, medical care one might be obliged to provide on their behalf.

Given the difficulties just outlined in discovering what should count as a canonical moral vision or understanding, it does not appear possible by sound rational argument to deliver a canonical content-full moral account of how one ought to proceed. For a sound rational argument to be conclusive for participants in controversies about how one should act in these matters, the participants must share the same fundamental premises as well as rules of moral evidence and inference. Individuals in substantive bioethical controversies are separated not merely by their conclusions, but often also by their moral premises (i.e., different thin theories of the good, different moral senses, different moral intuitions, etc.). As a consequence, the debates continue without a definitive closure, as one notes with respect to fundamental disputes regarding equality, abortion, and euthanasia. It does not appear possible in principle or in fact to draw moral authority from a single content-full account of moral rationality so as to justify one rather than another set of health care policies bearing on issues of equality of access, the availability of abortion, as well as the value one should give to human life.

Yet, there is still one expedient that can be invoked. If one is to resolve controversies in health care policy by means other than through a direct appeal to force (even if the force invoked is on behalf of a majority exerted on a minority through a democratic process), and all do not accede to the same religion or moral vision, and if sound rational argument cannot deliver a conclusion without first granting particular premises and rules of inference, authority can still be derived from the content of those who decide to participate in an endeavor. If public authority cannot be derived from force, God, or reason, it can be derived from actual persons.

It should then come as no surprise to find that those procedural mechanisms drawing their authority from the permission of those who participate are in fact the ones binding moral strangers across the world. Those are the ones that bind despite moral disagreements, and that function (at least in clandestine forms) even in the face of governments attempting coercively to impose a particular content-full morality by means of contracts and the market. This appeal to authorization can also justify a limited democracy in which the society it encompasses is not assumed to be equivalent to one moral community, with the consequence that citizens with different moral visions are at liberty to collaborate in a civil society that is morally neutral, as well as to create moral exclaves in particular collateral moral communities. In these circumstances, since the only source of authority is the permission of moral agents, it is persons, not humans as such, who are central. This approach to resolving moral controversies in the face of intractable moral difference establishes a sparse set of right- and wrong-making conditions: no value claims are advanced. Indeed, one does not even value the procedure of gaining authority. The possibility of gaining such authority functions rather as a kind of transcendental possibility.

Those who, as moral strangers, would wish to collaborate, can by permission give an authority to their joint endeavors: the authority of their concurrence. The limits on the authority of such joint endeavors will be the limits of the actual permission of those actually participating. Both these limits and the centrality of persons and their choices reflect the one source of authority in the absence of the guidance of God or of a particular notion of moral rationality. When moral strangers collaborate, insofar as they claim general secular moral authority, they will by default be libertarian. They will not be libertarian because they value liberty, but because the only source for generally secularly recognizable moral authority when moral strangers collaborate is permission. All who act within this practice will be able to advance generally secularly justifiable grounds for using punitive and defensive force. On the other hand, those outside will have no general secular grounds for protesting when visited with punitive and defensive force.

Those within the practice will understand the necessary conditions for the possibility of moral communication among moral strangers. These conditions will allow the identification of actions as right or wrong, but not as good or bad. Even the value of the practice itself cannot be made out in general secular terms. As with the market, each person who enters into the practice will enter with particular values: the price set is the outcome of a particular decision to trade. The price does not express an independently appropriate price, nor is there such a price to which one could make reference, though each of the participants may have a sense of such an appropriate price. The authority of the agreement establishes the authority of the price of the sale. There need not even be a generally agreed-upon value for the market. The same holds for the grounding framework itself and other practices that it can justify, such as limited democracy.

All of this is to say that, when we come to explore the challenge of discovering the value attributed to life, we do so in the context of a culture which has fragmented and is still fragmenting into competing moral narratives, and for which there is no generally accessible, rational, or transcendent glue by which to assemble the whole into one cohesive account. As a cultural circumstance, for some this may only be appreciated with difficulty in a Roman Catholic country such as Poland. However, the conceptual roots of the foundational moral difficulty are unavoidable. The only strategy available by which to secure general secular authority is to draw that authority not from a particular notion of rationality, right action, or the good, but rather from the permission of those who collaborate. This one generally justifiable source of secular authority is so sparse that it will authorize at most a limited democracy and will not establish a canonical value for human life. The character of this practice recognizes persons, moral agents, as cardinal, for they are the source of secular authority. In establishing a morality that can bind moral strangers in a secular world, one simultaneously draws a wedge between human biological life and that human life which is personal in the sense of being the life of moral agents, those entities who can participate in controversies and either agree or refuse to accept particular solutions.[6]

IV. POST-MODERNITY, THE GOSPEL OF LIFE, AND LIFE'S VALUE FOR MEDICINE

Given our circumstances, it will not be possible for secular bioethics to understand the evil in abortion, selective infanticide, third-party-assisted reproduction with donor gametes, physician-assisted suicide, or euthanasia. One will see taking place what is in fact transpiring: these practices will increasingly constitute taken-for-granted practices in secular health care. On the other hand, issues of free and informed consent, the right to refuse treatment, the recognition of moral diversity, and various moral analogues to the old American constitutional notion of rights to privacy, will become increasingly central.[7] When moral strangers meet, procedural morality takes center stage. Though one may attempt to preach the gospel of life, it will not be possible to bring others to its truth by sound, rational argument. Rather than decry the failure of philosophy, the pope of Rome would surely be better advised to reestablish the asceticism of the Fathers. Thus he could begin to sustain their narrative by drawing on transcendent roots while abandoning all hope that secular philosophy can disclose a rationally defensible content-full morality, as it is this hope, integral to the modern philosophical project, which beckons, but cannot be fulfilled. Indeed, it can only mislead (Engelhardt, 1995). Despite the history of this century and in the absence of a universal conversion to the true gospel of life, we will be left as moral strangers to frame limited endeavors, which will permit morally diverse actions, many of which we will have to tolerate, though they are very wrong.

Baylor Medical Center

REFERENCES

Abbot, Francis. *Report of Addresses at a Meeting Held in Boston, May 30, 1867, to Consider the Conditions, Wants, and Prospects of Free Religion in America*, Boston, 1868.

Cieszkowski, August von. *Prolegomena zur Historiosophie*, Berlin: Veit, 1838.

Engelhardt, H. T., Jr. *The Foundations of Bioethics*, 2nd ed., New York: Oxford University Press, 1996.

Engelhardt, H. T., Jr. "Moral Content, Tradition, and Grace: Rethinking the Possibility of a Christian Bioethics," 1995, *Christian Bioethics* **1**, 29–47.

Heidegger, Martin. *Grundbegriffe*, in *Gesamtausgabe*, vol. 51, Frankfurt/Main: Klostermann, 1981.

John Paul II. *Evangelium Vitae*, Vatican: Libreria Editrice Vaticana, 1995.

Kant, Immanuel. *Kritik der praktischen Vernunft*, in *Kants Werke*, vol. 5, Berlin: Walter de Gruyter, 1968.

Kant, Immanuel. *Immanuel Kant's Critique of Pure Reason*, trans. Norman Kemp Smith, London: Macmillan, 1964.
Kant, Immanuel. *Critique of Practical Reason*, trans. L. W. Beck, Indianapolis: Bobbs-Merrill, 1956.
Khushf, George. "The Sanctity of Life: A Literature Review," in *Sanctity of Life and Human Dignity*, ed. Kurt Bayertz, Dordrecht: Kluwer, 1996, pp. 293–310.
Kühne, Walter. *Graf August Cieszkowski*, Nendeln: Kraus, 1968.
Merleau-Ponty, Maurice. *Humanism and Terror*, trans. John O'Neill, Boston: Beacon Press, 1969.
Moore, Howard. *The Universal Kinship*, London: George Bell, 1906.
Nicodemus and Agapius, Sts. *The Rudder of the Orthodox Catholic Church*, New York: Luna Printing, 1983.
Taylor, J. *The Great Exemplar of Sanctity and Holy Life According to the Christian Institution*, printed by J. Flesher for Richard Royston, London, 1667.
Wildes, Kevin W. S. J. *The View from Somewhere*, in preparation.

NOTES

[1] Reports of the Christian proscription of abortion and of infanticide reach back to the *Didache* (II.2) and the Epistle of Barnabas (XIX.5), as well as to early canons of the Church. Already at the Council of Ancra in A.D. 315, canon 20 (Nicodemus and Agapius, 1983, p. 501) speaks to abortion. This condemnation is further underscored by canon 91 of the Quinisext Council (the Council in Trullo, A.D. 691, p. 395). As to issues of suicide, we find these addressed in some details by St. Timothy (+385), Pope of Alexandria, in his Question 14 (p. 898).

[2] In his First Critique, Kant argues: "This is the answer to the first of the two questions of pure reason that concern its practical interest: – *Do that through which thou becomest worthy to be happy*. The second question is: – If I so behave as not to be unworthy of happiness, may I hope thereby to obtain happiness? In answering this question we have to consider whether the principles of pure reason, which prescribe the law a priori, likewise connect this hope necessarily with it. ... But this [the distribution of happiness in exact proportion to morality] is possible only in the intelligible world, under a wise Author and Ruler. Such a Ruler, together with life in such a world, which we must regard as a future world, reason finds itself constrained to assume; otherwise it would have to regard the moral laws as empty figments of the brain, since without this postulate the necessary consequence which it itself connects with these laws could not follow," Kant, 1964, A809–B837, p. 638; A811–B839, p. 639.

[3] Kant, 1956, p. 133. "Die Lehre des Christenthums, wenn mann sie auch noch nicht als Religionslehre betrachtet, giebt in diesem Stücke einen Begriff des höchsten Guts (des Reichs Gottes), der allein der strengsten Forderung der praktischen Vernunft eine Gnüge thut." Kant, 1968, pp. 127–8.

[4] "Das des Selbstmordes noch zu übergehen scheint...." Kant, *Die Metaphysik der Sitten*, in *Kants Werke*, vol. 6, p. 425. Kant appears influenced by the strange Western Christian sentiment which led to placing masturbation as a sin equal to, if not worse than, adultery. See, for example, Thomas Aquinas, *Summa Theologica* II–II, Q 153–4.

[5] This lament is perhaps best articulated by John Paul II in *Evangelium Vitae* (1995). The pope of Rome observes: "In the background there is the profound crisis of culture, which generates skepticism in relation to the very foundations of knowledge and ethics, and which makes it

increasingly difficult to grasp clearly the meaning of what man is, the meaning of his rights and his duties" (p. 21).

[6] I have developed these arguments in greater detail elsewhere. See, for example, Engelhardt, 1996.

[7] The concept of rights to privacy as both legal and moral limits on governmental intrusions is found articulated in such American constitutional decisions as Griswold v. Connecticut, 381 U.S. 479, 85 S. Ct. 1678, 14 L. Ed.2d 510 (1965); Eisenstadt v. Baird, 405 U.S. 438, 92 S. Ct. 1029, 31 L. Ed.2d 349 (1972); and Roe v. Wade, 410 U.S. 113 (1973). This language has recently been rearticulated in the less congenial language of liberty interests. See In re Cruzan 58 LW 4916 (June 25, 1990).

ROBERT J. BARNET

CHANGING VALUES IN CONTEMPORARY MEDICINE

From the times of Hippocrates and Galen, medicine has held that the good of the patient is its primary concern. Throughout the centuries, there has also been a presumption that there are natural limits, that we dwell in community, in an imperfect world. Although not always evident in practice, medicine accepts the primacy of "do not harm," and implicitly recognizes the dignity of the human person. But the reality is not always all-inclusive. At times slaves were excluded, women set apart, and the marginalized ignored. Paternalism was dominant. It is that tradition that Hans-Martin Sass[1] describes as one in which the physician is "both acting professional and moral subject," with the patient's "moral virtues" limited to "compliance and hope." Although the Enlightenment had emphasized liberty and self-determination, its effect on medicine was minimal. Following the Second World War, there was a greater recognition of the importance of freedom of choice. A reappraisal both societal and professional, followed with a move in medicine to treat not only symptoms and disease, but to emphasize compassion and regard for the "whole patient." This was a challenge to the tradition that dominated the profession for over two millennia. However, strong paternalism still characterized most of medicine well into the 1960s.

With many asserting a virtual absolute right to self-determination, stronger patient autonomy did develop, in part as a reaction to a sense of restricted freedom. This trend was not unique to medicine; it was also central to the worldwide social and political changes of the past few decades. In those areas of the world where this is still not conspicuous, it is becoming increasingly evident. But, just as these new values emphasizing self-determination were close to being accepted as the norm, and as the scene was more coherent, new and different priorities appear which are displacing or at least threatening this new model. Values that are *different*, perhaps not unacceptable in themselves, are replacing the only recently emerging values. These newer values seriously challenge the character of both medicine and society. It is not that similar attitudes did not exist in the past, but that these, and the related social and economic structures, are more pervasive and more controlling. If the welfare of the individual is supplanted by self-serving entrepreneurial priorities because of this, then there is reason for concern, need for reflection and the necessity for reappraisal.

A.-T. Tymieniecka and Z. Zalewski (eds.), Analecta Husserliana LXIV, 209–221.

It would be one thing if only the status, power and control of professionals were challenged. If this most recent set of values served to reinforce respect for the dignity and liberty of the human person, there would be less reason for concern; but if the result is less appropriate goals and standards, the response should be quite different. If this most recent set of values diminishes the importance of patient welfare, then the problem is serious. The scene is complicated not only by the aforementioned assertion of greater patient autonomy, but also by increased individualism and a diminished sense of community. These changes characterize not only medicine, but society in general. Governmental and commercial entities have assumed greater control over activities traditionally rooted in community and family. At the same time there is a call for increased freedom in many areas of the world. Originally intended to benefit the individual, these changes established new dependencies, and left many less able to cope. This move for greater autonomy creates greater isolation and, ironically, often less freedom. With parallel developments in both medicine and society, patient-centered medicine, freedom to choose, and the social solidarity[2] required to implement these, become even more difficult to achieve. One option is to transfer the burdens of the current structures to those who are caught up within them and to make them *responsible*. An alternative is to seek radically different answers that challenge the present system.

HEALTH

There are diverse ideas about the proper role of medicine, as well as "health" in contemporary society. The World Health Organization defines health as an all-inclusive term encompassing virtually every aspect of human activity. In part because of this, there is a belief throughout much of society that there is a need for an all-inclusive package of products and services, defined and designated by the label of "health care," delivered and controlled by professionals. Current discussions about health include references to a "health care crisis" and a need for better "health care systems," involving the requirement for greater access to professional care with its technology and expertise. These are only part of the changes of the last fifty years. Advances in biomedical research and technical innovations are unprecedented. No more than five percent of the knowledge, techniques and resources that constitute our current medical armamentarium were available even as recently as forty years ago. In almost every area of the world, access to more sophisticated, technologically advanced and expensive treatment is seen as necessary, or at least desirable. These positions are reinforced by health professionals themselves who have bought into the new belief system. It is inappropriate to blame, as some do, patient autonomy for the current dilemma when the clamor for increased access to health care is largely the result of those who seek it, having accepted the claims and recommendations of professionals.

In developed industrial countries, chronic illness is the major reason for limitations and physical discomfort. In less developed countries, malnutrition and infectious disease, including typhoid, malaria, dysentery, cholera and now AIDS, are the important issues. A few simple, inexpensive interventions such as immunization, clean water, basic nutrition, simple antibiotics and intravenous fluids and the means to administer them, are often what is most appropriate, but most lacking. Worldwide, a limited number of people have access to the most basic health care benefits. These and other needs, including adequate housing, education, sanitation and programs directed at the elimination of violence, are often given lower priorities than sophisticated technology, medical and otherwise.

MEDICALIZATION

I characterize *medicalization* as a process by which certain activities, traditionally rooted in family and embedded in community, are redefined as appropriate for inclusion within the realm of professionalized health care. The defining guidelines assert that such activities as the care of the newborn and of the aged, our ordinary physical activities, choosing what we eat, how we cope with many illnesses and discomforts, and the rituals identified with dying, are, among others, the purview of certified health professionals. In characterizing this approach, Ivan Illich notes: "Medicine develops a system that creates a monopoly over the construction of bodily reality."[3] It is this medicalization which typically leads to the institutionalization and then the commercialization of traditional activities. It is a myth, however, that technology and the associated "advances" are either enabling or ennobling. As various aspects of social activity are medicalized, control is typically superimposed and extended. As professionals, we certify to such by creating a diagnosis to fit our claims as we stake out new territory. The diagnosis is critical since it signifies a claim, establishes control, and leads to dependency, entrapment and impotency. As members of the profession, even though we do not openly acknowledge the entrapment, it is we who are responsible. With freedom compromised, the autonomy the profession so recently resisted becomes its justification for abandonment when a cure is not possible. The move from the traditional long-term patient-physician relationship with its emphasis on trust, to the contemporary more fragmented pattern, often involving multiple physicians and a pattern of care involving specialization and sub-specialization, adds to patient isolation and uncertainty. A set of new needs and corresponding dependencies results. The structures to which the profession gave birth provide opportunities for exploitation. These new entities, however, are increasingly co-opted and controlled by power brokers and profiteers, some inside and many outside of the profession. The profession is enticed and then subdued by outside interests seeking an opportunity for profit. The resultant self-serving "deals with the devil" often exclude the professionals from decision-making and leave patients' best interests compromised.

Everyone has a moral claim on those goods of society that are necessary in order to function as a meaningful member of society. The implementation depends on the vision of what constitutes a "good society," on how the needs of others are met and on what resources are available. It is neither feasible, nor even desirable, for everyone to have unlimited access to all aspects of modern health care. It is not feasible because there are limits to our resources and there are other social needs that must be met. It is not desirable because of the entrapment, artificiality and inhumanity that is a part of modern health care, especially at the beginning and end of life. One result of the Enlightenment was to make social goods, such as health care when its time came, a matter of contractual obligations and political commitments. It is inappropriate for health care to be the sole or even primary goal of society. Priorities should be established in the context of what else must be sacrificed to attain them, both by individuals and society. It is not just an issue of what can be accomplished in terms of preservation and enhancement of functions or the prolongation of life, but also of the costs, financial, emotional or in terms of human dignity.

Modern societies, strongly influenced by an accent on material goods and a decline in religious traditions, have difficulty coping with death and finitude. The issue of physician-assisted suicide is a case in point. In the United States, at the present time, a majority (or close to it), of those surveyed, professional or otherwise, favor the legalization of physician-assisted suicide. There is significant acceptance of euthanasia in other countries, especially in some of the more industrialized ones. There are multiple reasons for this support, some societal and others related to the changing values and priorities of contemporary medicine and ultimately the understanding of the meaning of life. The fear of entrapment by de-personalized technological medicine creates an atmosphere that calls for an avenue of escape. Personal experiences, typically involving family members or friends, are cited to explain the rising support for assisted suicide. Liberation through euthanasia is seen as a preferred alternative to protracted pain and suffering on the one hand and technical entrapment on the other. No other choice is apparent. The possibility of finding meaning in pain and suffering is rejected; the possibility of continuing to find meaning in our lives is dismissed. Both would require the acknowledgement of our interdependence. This is a special problem, as we try to adapt in a society characterized by individualism with diverse and largely secular values. The burdens at the end of life generate insecurity. The patient, feeling useless and isolated, senses a need to "get out of the way," in part to minimize the "hardship" on others.

THE INROADS OF TECHNOLOGY

Medicine has shifted from being primarily a "healing art" to an endeavor shaped by technology and science. The power, control and priorities that the medical profession exert have changed with the shift from community-based, clinically oriented faculties, to institutional medicine often characterized by a concentration on research, technology and cures. But that is changing as well. These "institutions" are now often "without walls" and increasingly entrepreneurial. The power that physicians so recently held tenuously in hand is declining. Even for those in academic medicine, the milieu in which education and clinical practice are conducted is permeated with financial pressures, and dominated by limits, which are imposed, not chosen. These limits are established in the name of efficiency, typically directed at increasing profits or constraining costs, not by concern about patient good and better medicine.

Medicine has never been completely free of self-serving practices that place personal gain or scientific endeavors above the good of patients. The history of inappropriate experimentation by competent, often esteemed, physicians is well known. Here, in Kraków, the reminders of those experiences are too close for us to deny their reality. There have been disclosures about the use of potentially harmful radiation in the 1960s and 1970s without obtaining adequate consent. Fortunately such events are the exception, not the rule. But they continue. Recently, in North America, there have also been revelations about falsified data in breast cancer studies. Research often involves as subjects individuals who are vulnerable or marginalized, the very ones for whom there should be special concern. Often the rationalization for the compromise is that "others in the future" will benefit.

Pride and personal ego, rather than potential patient benefit, apparently were the primary motivation for some of the more recent deceptions. Nor is it always simple personal financial gain that results in compromises. It may be career opportunity or advancement, social standing, or peer approval. The profession is not immune to the influence of power, privilege and prestige. None of these goals are, in themselves, necessarily immoral. However, even staff or faculty appointments or professional advancement that are not based on merit or competence may result in compromised patient care, if not overt harm.

The dependence of physicians on high-tech life-prolonging interventions with the uncertainty and ambivalence about when it is no longer appropriate, complicates contemporary patient care, especially at the end of life. Fragmented, sub-specialized care adds to the disarray. These all add confusion to the decision about appropriate limits. Alternatives, such as hospices, are often excluded. The tendency for physicians to over-treat is bolstered by pressure from families who accept the claims of medicine that a favorable outcome is always possible. At times there is a "moral belief," on the part of both parties, that there is a need to prolong life. Those in medicine who cannot let go when treatment is excessively burdensome, or no longer effective, are often mesmerized by a technological (and at times, moral) imperative that seeks to maintain control to the very end. To do otherwise may acknowledge the fallibility of the profession. Both medicine and society, caught up in the emphasis on technology, deal inadequately with pain, and especially suffering. Compassion and meaningful relationships are often most wanting when their need is the greatest.

Too often, the vernacular or common sense ways of living rooted in the community are abandoned or forgotten. Many allow themselves to become dependent on technology and "experts." This is true even in those areas of the world where sophisticated health care is available to only a very few. This results in expectations, both societal and professional, that cannot be met. There must be either the acceptance of limits or the sacrifice of other priorities. Recent events in both France and Germany indicate that even those societies with a long tradition of providing virtually unlimited health benefits, are under pressure to limit benefits because of costly technology and medication.

ENTREPRENEURISM

A commercial or entrepreneurial enterprise focuses on results and profit. A moral endeavor focuses on respect for the human person and the needs of that person. A commercial enterprise operates with rules, objective standards of conduct and external controls. I propose a broader definition of "entrepreneurial" that includes not just direct financial gain, but also the previously mentioned triad of power, position and prestige. When medicine lets those elements determine its actions, it compromises its claim to being a profession. A profession has at its center a relationship based on trust; its guides for conduct are internal. The issues of integrity and virtue are integral and essential to any profession. The loss of that tradition diminishes the profession of medicine as well as the good of society. If medicine is a moral enterprise with the good of the patient as its goal, the answer will be quite different than if health is seen as a commodity and the prevention of death as the goal of medicine. If it is the latter, then medicine should be prepared to give up its privileges (if it has not done so already).

Medicine is combining with commercial forces to redefine its role. The growing secularization of both medicine and society loosens the connection with its ethical traditions and the importance of compassion. The dramatic events in Eastern Europe just a few years ago were the results of a combination of factors, including a desire for self-determination, but among them the desire of many for greater access to material goods. With such attitudes dominant in both society and medicine, abuse and exploitation are not unexpected. This new atmosphere coincides with the confusion and uncertainty about whether medicine is a business or a profession. Entrepreneurism within both the society and medicine, characterized by efficiency and productivity and reinforced by new goals of profit, exerts increasing control. Although the long tradition of beneficence and the importance of a compassionate and caring relationship in medicine have not completely disappeared, the increasing influence of entrepreneurial forces works in tandem with the reliance on technology. This results in greater outside control and increases the likelihood that the profession will be coopted. The integrity of the profession of medicine is under stress. In the United States there are excessive fees, unnecessary surgery and the medicalization and commercialization of such disciplines as nutrition and exercise. Investor-owned for-profit enterprises exert an ever greater influence on clinical medicine. It is not just the clinical practice of medicine but the academic and research branches that are influenced.

The experience in the United States is not unique. The impact of pharmaceutical companies on both practice and research is worldwide. In some countries, hospital appointments are decided by political connections or social standing. The scandals, both in Europe and Asia, of the distribution and sale of HIV tainted blood are well known. The use of bribes and gratuities to access government medical services is a recognized problem in countries with varying levels of resources. The marketing of organs for financial gain, a practice not unheard of, requires the collusion of physicians. In one Pacific rim country[4] it is conventional wisdom that rural practice is desirable because it provides a special opportunity to accumulate wealth. Financial incentives offered by drug companies include paid trips to medical meetings and, for the especially resourceful physicians, even free automobiles. In addition, there are opportunities to sell drugs samples to local "chemists" at significant profit. Direct harm often results from these practices because the desire to increase profit results in the use of adulterated, diluted, ineffective or mislabeled preparations. In another country, industrial interests have opened hospitals as for-profit ventures to provide testing facilities for biotechnology. The probability of conflict of interests in such situations is obviously great.

What reflections can I offer? First, there is a need to address whether the issue is simply one of freedom and the protection of rights or whether the issue is a deeper one that touches on the meaning of life. We can examine the meaning of life, beginning with an acknowledgement of the human condition and both its inherent richness and its fragility. People in many places and in other times had a great capacity to care for themselves and one another and could celebrate themselves the important aspects of life with meaning. This is especially true in moments of birth, when injured, while aging and in dying. We will suffer pain and impairments, and certainly die. Some will suffer greater pain, some more debilitating conditions, but all equally face death. There cannot be a return to a nostalgic and unattainable past; that is not a viable option. Developments such as the elimination of coronary heart disease and the elimination of cancer and AIDS, if they do occur, will leave the degenerative diseases prevalent. The elimination of all infirmity and the defeat of death is not possible. The challenge is to recognize those attitudes and practices that lead people to seek desperate answers and to find an alternative understanding. What is required is to explore the meaning of suffering and death, the appropriate role of medicine, and alternatives to what we do now.

MORTALITY, SUFFERING AND THE DENIAL OF DEATH

The denial of death is pervasive in contemporary society. The medical profession is not blameless, we hide behind euphemisms. The blind pursuit of health and cure, with its attempt to defeat rather than accept death when it is inevitable, is a serious disorder in itself. We often see death, even when inevitable, as failure. The profession first tends to over-treat, then, having failed to cure, often abandons the dying. This is due in part to the priorities and perspective of the profession, with the technological imperative being so dominant.

Viktor Frankl in *Man's Search for Meaning*[5] offers special insight. Frankl's ability to find meaning, in spite of the physical and psychological circumstances of Auschwitz, provides a unique perspective with which to approach the terminally ill. "In spite of all the enforced physical and mental primitiveness of the life in the concentration camp, it was possible for spiritual life to deepen. Sensitive people who were used to a rich intellectual life may have suffered much pain . . . but the damage done to their inner selves was less." Frankl found the isolation, and especially the denial of the prisoners' humanity, most telling. Not unlike the situation in which so many of the terminally ill find themselves, the majority of prisoners "suffered from a kind of inferiority complex." "We all had once been or had fancied ourselves to be 'somebody.' Now we were treated like complete nonentities. . . The average prisoner felt himself utterly degraded." The parallel with intensive care units and distant wards of nursing homes is frighteningly similar.

Of his own experience, Frankl writes:

> ... perhaps I was struggling to find *reason* for my sufferings, my slow dying. In a last violent protest against the hopelessness of imminent death, I sensed my spirit piercing through the enveloping gloom. I felt it transcend that hopeless, meaningless world, and from somewhere I heard a victorious "Yes" in answer to my question of an ultimate purpose.

Frankl continues:

> Fundamentally... any man can, even under such circumstances, decide what shall become of him, mentally and spiritually. He may retain his human dignity even in a concentration camp. Dostoevski said once, "There is only one thing I dread: not to be worthy of my sufferings." These words came frequently to my mind after I became acquainted with those martyrs whose behavior in camp, whose suffering and death, bore witness to the fact that the last inner freedom cannot be lost. It can be said that they were worthy of their sufferings; the way they bore their suffering was a genuine inner achievement. It is this spiritual freedom – which cannot be taken away – that makes life meaningful and purposeful.

If Frankl, and those who were with him, could find such insight under those circumstances, is it not possible that modern society and medicine are missing something essential? Philip Aries,[6] The French historian, writes of this "tame death" of another age, one that was "tolerable and familiar, affirmative of the bonds of community and social solidarity, expected with certainty and accepted without crippling fear." He contrasts that "tame death" with the "wild death" of contemporary technological medicine which so often is of no real benefit to those suffering.

It is not surprising that suffering in the face of death involves isolation and a sense of helplessness. The stress on autonomy and its claim that we totally control our own bodies, although seemingly a reaction to the technological imperative, actually works in tandem with it. In acquiescing to the power of technology, at the same time as we stress self-determination, we also create further isolation. There is a need to re-examine the concept of community, to explore the notion of solidarity, to recognize our interdependence and the importance relationships have in relieving suffering or finding meaning in it. As is implicit in Frankl, to simply be present with one who is suffering offers an opportunity to confirm the importance of a loving relationship. It allows the sufferer to be affirmed, less isolated. But that need must first be recognized and acknowledged.

Frankl argues that human life, under any circumstances, never ceases to be without meaning, and that the hopelessness of our struggle does not detract from its dignity and meaning. "[S]omeone looks down on each of us in difficult hours – a friend, a wife, somebody alive or dead, or a God – and he would not expect us to disappoint him. He would hope to find us suffering proudly – not miserably – knowing how to die." Alternatives to the institutionalization and professionalization of the end of life should be nourished. Ivan Illich observes:

> Such an idea – to teach people how to suffer and die in this society – is shameful. Care for others through professional bereavement counseling, specialized education for dying ... aim directly at the destruction of virtue, at the obliteration of the traditional arts of suffering and dying. Each culture develops such practices over hundreds of years and, in the West, we can still find vestiges of them, especially among marginal people.[7]

It is a tragedy that we preclude the possibility of people finding meaning in their illness as they approach death. Support for assisted suicide draws support from individuals who have rejected the prospect of dying in a medically managed setting, whether in hospital or out. They are denied the opportunity that people need to find themselves and to establish or affirm their relationships with others. What is needed is for people to be able to die, held by friends and family, in comfort *and with meaning* in familiar surroundings. That option needs to be more available, in the setting of home and community as well as in hospices.

Frankl's experience and observations can serve as a guide for those in health care, by encouraging them to give affirmation because it is part of our role both as fellow humans and as professionals. In this way, the caregiver is affirmed as well. In this way, neither the lives of the one who dies nor of those who know the person are diminished. Too often, their choices, even the simple ones, can be meaningful. There must be another course, something other than a struggling, isolated soul locked in combat with the mechanical robot of tubes and wires. There must be something other than an "existence" defined by the medical bureaucracies of specialists and health care teams; something other than total isolation and the loss of meaning associated with absolute autonomy. The end of life too often takes place in an environment that fails to allow patients to find meaning in suffering or in death.

There is a need to re-evaluate the role of medicine and its goals of preserving life and relieving suffering. It is important to distinguish between pain and suffering. Pain is the body's response to disease or injury; suffering is our response to the pain. There are no technological solutions for all problems. We cannot eliminate all pain, cure all disorders or avoid death. As part of the answer, we need to demystify life and death and to encourage people to seek out the simpler, the vernacular, in dealing with both. Flannery O'Connor in *The Habit of Being*[8] notes that "illness is a place; it is where one resides at a time of crisis." If we take such a perspective, then what is needed is the company of those who are important to us, not a technological fix. That is especially true in terminal illness. In Tolstoy's *The Death of Ivan Illych*,[9] the suffering of Illych came not from the illness or the pain, but from the fear of dying and the isolation, especially from his wife and son. The Doctor offered to relieve his suffering, but did not understand the difference between pain and suffering. Tolstoy writes: "The doctor said his physical agony was dreadful, and that was true; but even more dreadful was his moral agony, and it was this that tormented him most."[10] It was the company of the simple servant, Gerasim, that made a difference in those final days. For he saw death as an expected reality and Illych as a person. Then Illych was able to own the illness as part of his experience, and not feel separate from it. We learn that death is *part* of life. It was the recognition that death was the fulfillment, the acknowledgement, the conclusion of a life, that became the opportunity to reflect on and to celebrate his life. And in his last few hours Illych was able to reach out to his son and wife and set aside his dread that life was empty and without meaning. As was the turning point with Illych, it is often that recognition of the vulnerability and mortality of the human person, with a dignity and uniqueness independent of all the frailties, a recognition of the human condition, that is crucial.

In *The Troubled Dream of Life*,[11] the American philosopher Daniel Callahan suggests that modern medicine extends the role of the technological imperative beyond illness and disease and attempts to bring all aspects of our life, and even death, under the mantle of medicine and science. He maintains that "medicine must abandon the cultic myth that in the cure of disease lies the cure of death." But where does that leave us, without an alternative or a deeper meaning in life? This appraisal requires an exploration of the question of whether medicine is to continue as a profession and a moral endeavor, or whether it is to become a self-serving commercial enterprise. The outcome will be contingent on finding a meaning in life that acknowledges pain and suffering. It is not just those in the profession of medicine who must address the issue, but also society, which must explore how to approach living, disease, disability, suffering and dying.

CONCLUSION

A fresh perspective is needed, one that retains the dignity of the human person as central, recognizes the fragility and contingency of the human condition, the reality of limits, and that cultivates the art of living, the art of suffering and the art of dying. What is required is a recognition of a meaning of life that acknowledges the human condition and its inherent richness. It is my claim that this can best be accomplished in a milieu that stresses the importance of community, the notions of responsibility, solidarity and interdependence, and that seeks to find ways to develop and nourish relationships which play a special role in easing or finding meaning in suffering, while at the same time recognizing the dignity of each and every human person.

Whether our interest is in the philosophy of medicine, clinical medicine, research or education, we need to step back and ask if our goals and actions place that dignity of the human person, our relationships, and the welfare of the individuals, in a place of primacy. How can that be implemented? Each step of the way, with each encounter, in the classroom, clinic, or across the table as we break bread together, we can consciously try to recognize that there *is* a fundamental worth and dignity in each person we encounter. If we could only somehow visualize that reality in our daily activities, and act with that in mind, we would make a beginning. We can do this as we attempt to reject an unjustified dependence on technology and systems and replace those entrapments with a recognition of our own interdependence, not just with each other, but with our communities and our environment. If this cannot be done, both the profession of medicine and society will have failed.

Reno, Nevada

NOTES

[1] Sass, Hans-Martin, "The New Triad: Responsibility, Solidarity and Subsidiarity," *The Journal of Medicine and Philosophy*, Vol. **20**, No. 6, Dec. 1995, p. 588.

[2] Solidarity is defined as "the willingness to share the risks of human existence with other persons or groups" (Ter Meulen, H. J. Ruud, "Limiting Solidarity in the Netherlands: A Two-Tier System on the Way," *The Journal of Medicine and Philosophy*, Vol. **20**, No. 6, Dec. 1995 p. 615). I prefer an understanding that identifies the concept as one in which, in the context of community, there is a recognition of our relationship with others, our interdependence, and that we are all entitled to share in the goods and resources of society so that our basic needs are met and we are able live with dignity.

[3] Illich, Ivan, *Blasphemy, A Radical Critique of Technological Culture*, circa 1992 (unpublished).

[4] Personal communication.

[5] Frankl, Viktor E. *Man's Search for Meaning*, Boston: Washington Square Press, 1984.

[6] Aries, Philip, *The Hour of Our Death*, New York: Oxford University Press, 1981.

[7] Illich, Ivan, *The Empty Sacrum*, circa 1992 (unpublished).

[8] *The Habit of Being: Letters of Flannery O'Connor*, New York: Farrar, Strauss, Giroux, 1979, p. 320.

[9] Tolstoy, L. N., *The Death of Ivan Illych*, translated by Lynn Solotaroff, Bantam Books, 1981.

[10] Ibid., p. 126.

[11] Callahan, D., *The Troubled Dream of Life*, Boston: Simon and Shuster, 1993.

DONALD EVANS

CLINICAL DECISIONS, IMPAIRED CAPACITY AND THE PUBLIC INTEREST

INTRODUCTION

I am concerned with clarifying the notion of clinical decision-making by looking for the boundaries of the notion. To do this I propose to examine some cases which I perceive to occur at and beyond the limits, and to mark out the differences between them. What is said of these cases and what consequences flow from our description of them may well have a considerable bearing upon cases which fall safely within the boundaries of proper clinical decision-making. The cases all involve the impairment of capacity of the patients to participate in decisions regarding their treatment. I shall not describe the unobjectionable cases as central cases because such a description might suggest a security about our knowledge of what actions it would be appropriate to take where such impairment was identified. We might, for one reason or another, be uneasy about clinical decisions made in cases at any point in the spectrum of patients with impaired capacity.

The cases I have in mind all involve situations where we are dealing with decisions involving compulsory treatment of persons for purportedly clinical reasons. I use the word treatment in its broadest possible sense here, including removal from place of abode or work, restriction of freedom of movement or liberties in general, detention, medical examination and therapy. The decisions with which I am concerned will be made in the interests of persons other than the persons so treated. I shall compare decisions made in the contexts of public health protection and mental health as properly involving the ascription of impaired capacity to consent, to those compulsorily treated. I shall contrast these with a recent case of the treatment of mothers suspected of life-threatening child abuse in the context of Munchausen syndrome by proxy.

THE LIMITS OF IMPAIRED CAPACITY

At first it might appear that the notion of impaired capacity to participate in the process of decision-making is equally under strain in all these cases – even the mental health cases when the dominant consideration is the dangerousness of the patient and the threat posed by that patient to the health

A.-T. Tymieniecka and Z. Zalewski (eds.), Analecta Husserliana LXIV, 223–238.

and welfare of others. However, a little reflection will show that wherever any coercion is involved in the treatment of persons, whether as the subjects of violent assault or threat or where some compulsory measure is applied to them, then their capacity to affect what is done to them is impaired. Most obviously, in the field of public health protection, this is due to the explicitly external constraint of the law which steps in and enforces measures such as compulsory removal to hospital and compulsory detention. The citizens subjected to such measures are, in fact, deprived of the legal right to determine what shall be done to them. Insofar as this is true, then their capacity to influence decisions concerning their treatment is grossly impaired.

A more subtle case could be made for regarding persons so involved as having impaired capacity. This case would be made in terms of internal rather than external constraints. These constraints may be seen at two levels, one being more fundamental than the other.

First, it may be argued that the possibility of an objective and impartial judgment concerning one's own treatment is necessarily threatened when the decision concerns precisely whether the interests of others should outweigh one's own interests in the matter. A determination by oneself of what should be done to oneself by others is normally one which is couched in terms of what it is in one's own interests to allow others to do to oneself. This is typically the way in which competent patients provide proper consent for procedures. An inability to assess what is in one's interests is a criterion of the incompetence to make such a decision. But in the area with which we are now concerned, the decision does not devolve on what is in the interests of the patient, but rather on what may be in tension with what are his own interests. Thus, seeing clearly what are one's own interests is precisely not what is necessary for a responsible decision to be made in the context of public health protection, where one's own condition is what constitutes the threat to the public. It may therefore be thought that the inevitable tension between one's own interests and those of others adds a problematic dimension to decision-making by the patient in this context. Thus, it may be argued, there will always be the threat of improper considerations (one's own interests) clouding the issue.

The second, and more fundamental, level takes note of the fact that some patients may be sufficient to resist the pressure which inevitably arises from their own interests. It may be felt, however, that seeking the consent of the patient is irrelevant in this context. That is, obtaining consent from the patient would lead to the application of the measures which restrict the liberties of the patient but, equally, failure to obtain consent would lead to the imposition of the same measures. This would make a nonsense of the whole procedure of seeking consent. The interests of others must prevail in this context; thus, seeking consent becomes an idle wheel. It is a matter of logical rather than psychological necessity, therefore, that the capacity of the patient to determine what decision should be taken is empty. That is, even if a given patient is able to weigh responsibly his interests against those of others in this setting, it would be a redundant capacity. His ability to influence the outcome of a proper decision-making process would be reduced to nothing.

PUBLIC HEALTH PROTECTION AND IMPAIRED CAPACITY

There are two major considerations crucial to determining the acceptability of any restriction of what can be regarded as the reasonable expectations of liberty of citizens. The first concerns the nature and degree of the threat constituted by the person in question – in the case of public health protection this will involve identifying the disease conditions which need to be specified as notifiable. The second concerns the nature and degree of the sanctions imposed on individuals, for example, the carriers or sufferers of the condition in question, and the reasonability of the grounds of suspicion that those individuals constitute a danger, for example that they may be sufferers from or carriers of such diseases. That is, justification must be afforded for any restriction of the liberties of citizens. To put the matter in terms of impairment of capacity, any measure which will constitute the construction of impaired capacity in the citizen to determine how he or she shall be treated must only be adopted where there are good reasons which have nothing to do with the state of mind of the person to be treated.

It is important to try to determine the parameters of such justifications in the health care setting. They will range from the commonplace, which may well be overlooked, to the rarer but more troubling cases. The greater number of such constructions occur in the context of resource allocation in health care. Whether a patient is offered treatment, or a given treatment, or a choice of treatments, will be a matter which is beyond that patient's power to determine. This can be true where the patient is willing to purchase a service, but is certainly the case in the systems of state provision. There, treatment is often made unavailable due to the pressure of demand and the relative need or convenience of other patients. Those so denied are often kept in ignorance of the decision, or at least of the reasons for the decision. For example, through use of procrastinating techniques such as waiting lists, patients might believe that they will eventually be allocated treatment when there is, in fact, no intention of treating them. In these cases and in cases of blank refusal, good reasons for the failure to provide treatment should be forthcoming.

The more troubling end of the spectrum of cases of constructed impairment of capacity to affect treatment decisions involves the question of whether the construction occurs in the pursuit of health care goals, that is, cases where it is doubtful whether it is the responsibility of the health care professional to restrict the reasonable expectations of liberty of the patient. For example, whereas the spread of infectious diseases in the population is clearly a matter of clinical concern, insofar as clinical expertise is necessary to detect the threat and determine how it is best averted, it is not obvious that all threats to the health of citizens are a matter of clinical responsibility. What about cases of physical injury? Certainly clinicians may be expert at detecting and treating such conditions, but is it their responsibility to protect people from such injury? They may warn authorities of the dire perils of brain damage resulting from motorcycle crashes, and the advisability of wearing crash helmets, but it is not their responsibility to enforce such measures. That responsibility lies elsewhere, as does the responsibility to protect people from injuries inflicted by others. Matters are even more doubtful when the concept of health is cast more widely than physical well-being – as occurred in some alleged cases of the abuse of psychiatry to protect citizens from ideological harm in the former Soviet Union.

If we consider the question of the nature of the threat to others, we may be puzzled by the selection of diseases covered by public health legislation. For example, should the seriousness of the prognosis of the condition specified be the main criterion? Or should it be the likelihood of infection of others? It could hardly be the latter – at least as a stand alone criterion. Influenza is highly infectious but would not normally figure in any list of notifiable diseases. The common cold would be an even better example. The likelihood of infection is important, but only when linked to the seriousness of the consequences of suffering the disease in question. If the disease is fatal in many, most or even all cases of a given infectious disease condition, this would present a prima facie case for its inclusion in any list of notifiable diseases. Yet acquired immune deficiency syndrome only appears in certain sections of the Public Health (control of diseases) Act in England and Wales; and syphilis, even when there was no known treatment for the condition, never figured as a notifiable disease. Perhaps the reason for the restricted appearance of such conditions concerns the unlikelihood of infection from the patient, and neither criterion is sufficient as a stand alone justification for the listing of any disease as notifiable.

But how likely must the possibility of infection be? And how serious the consequences of infection? These are difficult questions to answer, and tensions occur when different examples are compared. For instance, if the degree of possibility of infection is very high but the range of risk in terms of numbers of possible contacts is small, should this attract the sanctions envisaged in the law? To take a case in point, the positive HIV status of a husband puts his wife in very considerable danger of infection, and, moreover, of an infection which carries with it a dreadful prognosis. But is the relatively limited range of possible contacts offering the kind of setting necessary for the transmission of the disease sufficient to keep the disease off the list of notifiable diseases? It would appear that this is at least part of the explanation for the absence of the condition from the list applied to most sections of the Act in England and Wales.

Section 36 allows that a person not suffering with a notifiable disease but carrying an organism capable of causing it may, in the interests of their families, be medically examined by order of a magistrate. One criterion canvassed here is precisely the likelihood of transmission of the organism. Even so, however likely such transmission would be in the case of the HIV husband, this section of the Act would not make it an offence for him to expose his wife to the infection, as acquired immune deficiency syndrome is not a disease to be notified of in this section of the Act.

Could the wife be protected by Section 17, which makes it an offence for a person knowing that he is suffering from a notifiable disease, to expose other persons to the risk of infection by his presence or conduct? The answer must be negative, as the exposure has to occur in either a street, public place, place of entertainment or assembly, club, hotel, inn or shop. In any case it will be argued that to have positive HIV status is not to be suffering from acquired immune deficiency syndrome, which, in any case, is not a disease covered by this section of the Act.

There is just one possibility of protection for the wife in the Public Health (Control of Disease) Act: That is Section 38, which concerns the possible detention in hospital of persons suffering from a notifiable disease. AIDS does come under the aegis of this section. However, once again it could be argued that the HIV-positive husband is not in fact suffering from the infectious disease of AIDS. In any case he would first need to be hospitalised before such a detention was possible. It would be possible under Section 37 for such removal to occur where the husband was suffering from AIDS. But, as we have noted, this would not apply to the husband who simply had positive HIV status. Even so, it should be noted, there has only ever been one application to the courts for detention under this section of a patient suffering from AIDS (Manchester 1985). This was granted but led to great public discussion, which is presented as a powerful disincentive for Public Health officials to seek such orders.[1]

Thus it is not clear that the severity of the risk to others or the likelihood of the causation of harm to others alone entail an impairment of the competence of patients to decide what shall or shall not be done to them in terms of public health legislation. Presumably what is missing in the HIV case is the lack of wide-ranging possibilities of infection – even though the word "epidemic" does often occur in discussions of how the phenomenon should be handled, and of the character of the threat it poses to populations. The necessity of blood-to-blood contact dramatically delimits the range of possible infection by an individual. One diphtheria or tuberculosis sufferer in a crowded cinema obviously constitutes a greater threat than a highly malevolent HIV carrier in the audience. (I here set aside other possible explanations of the exclusion of acquired immune deficiency syndrome from many sections of the Public Health (Control of Diseases) Act, such as the political sensitivity of the issue.)

THE MENTAL HEALTH ACT AND CONSTRUCTED IMPAIRMENT OF CAPACITY

If we compare cases of imposed restrictions on persons under the protection of public health measures and under the Mental Health Act in England and Wales, however, we see that the range of persons open to harm is not a necessary consideration for producing impairments in the capacity of people to determine what shall or shall not be done to them. If the sufferer of a mental illness constitutes a danger to the health or safety of others, and the relevant others could be as few as one, then it is possible under the Act to remove that person to hospital and detain him. It may be that the chances of harm are no higher than in the case of the HIV positive husband, though usually the dangers will be of a more dramatic and visible nature.

Why such a person should be subject to impairment of capacity to affect the "treatment" decision concerning him, whereas the HIV positive husband need not be, is a question worth asking, though it will not be my task to pursue it at length here. Is it that there is a presumption in the case of the mentally healthy individuals that any curtailment of their liberties automatically calls for a justification whereas this is not so with the mentally ill? Is it that our perceptions of the status and rights of people suffering mental illness are radically different from those of mentally healthy individuals? If so, is this difference justified? It is certainly not explicit, but plausibly it can be said to be implicit in the quite different attitudes we take to the production of impairment of the capacities of the groups to influence their "treatment" decisions. Remember that we are here interested only in the cases where the impairment is related to the interests of others, and not where it is related to some psychological incapacity.

Certainly the Mental Health Act allows that the criterion of danger to others alone constitutes a sufficient reason for the restriction of the liberties of those suffering from mental illness. We need not raise the question of the competence of the patient to make decisions at this point, as this does not need to come into play in the production of impaired capacity to determine these particular treatment decisions. Nevertheless, I shall suggest that it is the running together of the two senses of impairment of capacity, constructed and otherwise, which facilitates a greater willingness to remove patients from the process of decision-making with respect to their treatment under the Mental Health Act than under the Public Health (Control of Diseases) Act. It is further possible that we are prepared to subsume certain patterns of offensive behaviour under the umbrella of mental health problems in order to manage them more readily. Here doctors and health carers more generally may be press-ganged into the business of social control, and placed in an invidious position where the normal presumptions of the carer-patient relationship are thrown into confusion.

The central feature shared by the mental health cases and the public health cases we have considered is that the person who is to suffer constructed impaired capacity is likely to engage in behaviour which is disapproved of because of the threat that behaviour represents to the interests of others. What distinguishes the mental health cases from the public health cases? It is simply that the persons whose capacity is constructively impaired in the former cases must be suffering from a mental illness or a disorder of the mind. It is not by virtue of that illness that the constructed impairment of capacity occurs. Apart from the threat posed to the interests of others, the person would not suffer such constructed impairment. In this connection, note that the mental illness might not be sufficient reason for sectioning under the Act, but the added feature of dangerousness to others provides a justification. It might be asserted that suffering a mental illness necessarily involves some reduction in the ability of the patient to responsibly determine how he or she ought to behave. Thus, when some element of danger to others arises, the constructed impairment of capacity only extends what is already identifiable as impaired capacity to determine how one shall behave, without the dangerousness element being present. Whilst certain mental health conditions may plausibly be so regarded, it is uncertain that the plausibility stretches across the whole spectrum of conditions which mental health professionals are concerned with in practice. There has been great discussion about whether mental illness should constitute an excuse for unacceptable and anti-social behaviour. The debate has centred around the question of whether the medical model, or more generally an illness model, is a proper one in this area. I am now suggesting that we should not jump too hastily to one or the other side of this polarised debate. Without taking the extreme position adopted by Thomas Szasz,[2] who doubts that the description of any mental condition as an illness is ever proper, we may enquire as to whether the states referred to in the Mental Health Act as mental disorders not involving mental illness nor mental impairment do provide us with an identifiable impairment of capacity, irrespective of that constructed by others who wish to contain the behaviours of the persons in question.

MENTAL DISORDER AND IMPAIRED CAPACITY

What is it which distinguishes the latter cases of constructed impairment involving mental disorder from those which simply involve the protection of public health? Personality disorders may amount to no more than that the person in question is likely to be involved in syndromes of behaviour of which people generally disapprove – viz, anti-social behaviour. Clinicians may be able to offer no explanations for these behaviour patterns. The disturbance of mind is entirely constituted by the occurrence or likely occurrence of this behaviour. The behaviour or the behaviour patterns are constitutive of the disorder. Are we entitled to group these cases together uncritically with cases of mental illness where we feel that the perpetrators of unacceptable behaviour are not responsible for their actions because of a psychological deficit, which takes the form of an incapacity to make competent decisions about their own behaviour? The latter might be said to suffer an impairment of capacity which is not constructed on the basis of our disapproval of the behaviour patterns which they are likely to exhibit, but rather on grounds of identifiable states of mind which can serve as explanations of those patterns and which can be identified independently of them. These will be patterns over which the actor may be said to have no, or at least reduced, control. I have in mind, for example, some forms of abnormal personality which involve compulsive behaviour. These will often involve phobic states where the person suffers persecutory feelings which are irrational, at least in dimension. For example, they may fear being poisoned by dirt or attacked by some article which has gotten out of place. Such feelings result in obsessive behaviour like constant washing of the hands or tidying up, when there is little dirt or untidiness about.[3] Similar behaviour patterns may be exhibited in Alzheimer's disease where repetitive activities, like the constant filling and emptying of a shopping basket, relate irrationally to familiar activities which once played a proper role in the life of the sufferer. In the latter case we can identify brain states, independently of the behaviour, which serve as explanations of the behaviour. The state does not reduce without remainder to the behaviour. It may not be as clear that we have such an appeal in the obsessive behaviour cases where there is no identifiable corresponding brain state. Yet we should wish to distinguish those behaviours based on palpably false but unshakeable beliefs and such unreasonable fears, from other behaviours which, though we may find them undesirable, cannot be so explained.

We can contrast the above-mentioned with cases where abnormal personality is ascribed to some people exhibiting obsessional behaviour, as though the description also served as an explanation of the behaviour. It would take the form of describing the person as having an obsessional personality. The description of the personality is simply a description of the behaviours of the person. It seems that the condition reduces without remainder to the behaviour patterns concerned. Indeed, the personality disorders are in general defined as being disorders of character where character is defined as: "the acquired pattern of habits, attitudes and ideals which render a person's actions stable and predictable."[4] Behaviour which is socially deviant, that is, behaviour of which we generally strongly disapprove, has been said to arise from this source and has been regarded as "moral insanity" since 1835. But no defect in reasoning or intellect, no insane illusion or hallucination is involved. Any one of these considerations might tempt us to attribute impaired capacity to a person in order to control what he does. However, we have, in the Mental Health Act, constructed impaired capacity in individuals so behaving without any of these features, on the grounds of the "seriously irresponsible" nature of the conduct.[5] Indeed, where some underlying brain damage is responsible for such behaviour, in an indisputably clinical case, then the case is not regarded as one of psychopathy but one of pseudo-psychopathy.[6]

Clearly, the element of danger to others plays an important role in constructed impairment of capacity in these cases. Such impairment only occurs in cases of sociopathy when the disorder is of the aggressive type. In the Mental Health Act, the danger of self-harm can form a justification for the construction of impaired capacity where the person is suffering from a mental illness.[7] But it is no longer believed, even if it once was, that suicidal behaviour necessarily denotes mental illness. It might be a sign of such a condition, but the condition cannot reduce to it. Some independent feature of the person must be identified as constitutive of the illness. Why then should it be different in the cases of harm threatened to others? In the absence of such independent data, there is no justification for considering that the actor needs the administration of some kind of health care intervention. Granted, the behaviour of the persons concerned may be deviant, even grossly so in the sense of greatly deviant from anything which approximates to a norm of human behaviour. Why are we tempted to call such deviance "sick" when it refers to behaviour which is also morally gross, but not when it refers to extraordinary levels of kindness or saintliness? That very few people manage to live like saints does not justify some moderating health care intervention to curb the behaviour of those who do so live. Here one can only be impressed with the claim that we have devised a sterilised form of social control in the name of medicine.

Indeed, we do not have to resort to extreme examples of deviant behaviour to make the point. How grey and uninteresting life would be, but for the small minority of exhibitionist personalities who offer us so much entertainment and interest in theatre, sports, politics and the media, where the behaviour can range from the unashamed baring of the body in public places to the baring of the innermost soul to millions of spectators. Where does the cut-off point come between the abnormal and such traits of personality as are manifested in such people? Indeed other traits, such as aggression, are to be found in almost everybody to some degree. It has been argued that there is neither a satisfactory theoretical account in medicine of the distinction between traits of personality and personality disorders, nor any demonstrable link between such syndromes of behaviour as we have cited and fundamental abnormalities of a histological, biochemical or molecular kind.[8]

We must, therefore, be careful to resist the temptation to uncritically extend the model of impaired capacity to other cases of unacceptable behaviour where, for purposes of social convenience or control, we construct an impaired capacity to determine treatment decisions in order to facilitate public protections. That is not to say that there should not be public protections from such behaviour, but simply that it is not the business of medicine to be providing them.

We have noted thus far that the situation as it currently stands in the law of England and Wales allows for an imposed impairment of capacity to determine treatment decisions in the interests of others in two areas, viz. public health protection and mental health. Thus the *de facto* situation is that some justification for such imposed impairment is thought to be proper. However, we have seen that the line is not easily drawn between cases where impaired capacity is properly constructed by health carers, and those where it is not. Let us look at a given set of cases where this vagueness is particularly troublesome. Maybe it will simply serve as a reminder that the line is inevitably vague, maybe it will help construct some guidelines as to where such a line should be drawn.

A TEST CASE OF CONSTRUCTED IMPAIRED CAPACITY

To illustrate the borderline nature of some of the ostensibly justifiable cases referred to above, I now propose to compare them with a recent group of clinical cases which bear considerable similarities to them but which, I contend, differ from them in sufficient respects as to place them beyond the limit of justifiable constructions of impaired capacity.

I have in mind the removal to hospital of mother and child where the mother is suspected of inflicting life-threatening abuse on the child.[9] The nature of the alleged abuse is rather special. It is inflicted on the child for the purpose of attracting the attention of health care professionals. It may be that it is a development of the creation of fictitious signs of a painful kind. The case in question involves the production of apparent signs of disease in the child – most commonly in the form of apnoeic attacks. In other cases startling loss of weight, blood in the faeces, rashes and abnormal urine have been engineered. The title of *Munchausen syndrome by proxy* has been conferred on such cases as they are seen to illustrate an extension from Munchausen syndrome itself, which consists of the invention or production of false signs of disease by a patient in herself in order to gain the attention of health care professionals.

The treatment, in the broad sense, of the mother and child is as follows: The mother is told that the child needs to be admitted to hospital for further tests or observations. She is invited, indeed persuaded to accompany the child. If any reluctance is shown, then considerable pressures are applied.[10] The move is made ostensibly to treat the child, but the mother is made to feel obliged to remain with the child throughout. The two are kept together – the child in the cot or bed as the patient, and the mother as an accompanying "caring" parent. An around-the-clock watch is kept on them, both by health carers and by means of covert video surveillance. Protecting the child is not the point of the video surveillance, as standard surveillance achieves this and has to accompany the video taping in any case. The videotaping is designed specifically to produce incontrovertible evidence of attempted suffocation of the child by the mother.

Once an apparent attack occurs, the health carers intervene, though enough time (about twenty seconds) is allowed to elapse for adequate evidence to be recorded. This time is thought to be allowable as no permanent damage by suffocation could occur in such a short span. The tape is then viewed, though not with the mother's knowledge. If the evidence is not thought to be conclusive, then the situation is allowed to continue until a further episode – if there is one – is recorded. Such surveillance has proceeded for periods of up to fifteen days. It has also been claimed by one mother so treated that she was not allowed to hold the baby during the whole of the surveillance. (See reference to the Gwynneth Tenney case.) Health carers visit the couple periodically as is normal in the hospital setting.

It is my contention that the mothers in these cases suffer constructed impairment of capacity to determine their treatment. Though the removal to hospital is not compulsory, their subjection to covert video surveillance in the name of providing medical care to the child is a limitation of their reasonable expectations of liberty. There is no evidence of what would happen if the mother refused absolutely to accompany the child. It is probable that an emergency protection order which can be issued on an *ex parte* basis, separating the child from the mother, would be enacted by the courts. Such an order could be followed by an interim care order maintaining that separation. The mother is thus placed in a Catch 22 situation. If the procedure is cast in terms of examining the mother for the factitious disorder of Munchausen syndrome by proxy, then she is actually being observed for diagnostic purposes without her consent. I will canvass reasons for saying that the latter is not, in any case, a proper description of what occurs.

AN EVALUATION OF THE CHILD ABUSE CASE

We may wish to question whether the paediatric team employing this technique is providing clinical care to the child? The only elements of practice possibly involved, as we have noted, are diagnostic activity and preventative activity.

Consider first the diagnostic activity. It is rather strange. No treatment is offered the child unless it suffers an attempted suffocation. But in this case, the treatment is not for a condition suffered by the child at hospitalisation. Indeed, it is only for injuries suffered after hospitalisation and by courtesy of the contrived environment offered by the hospital. The team admits that this setting increases the likelihood of such injury being enacted on the child.[11] "Parents were more likely to impose upper airway obstruction when, due to close nursing observation, they had been unable to demonstrate their child's symptoms for a period of time." There is one court judgement of a case where the apnoeic attack suffered before hospitalisation was agreed to be spontaneous and where imposed apnoea was produced by the mother for the first time in the context of the hospitalisation of mother and child.[12] The team is almost cast in the role of *agent provocateur* in such cases, creating the legal dangers of entrapment. If there is an element of diagnosis involved with respect to the child, it is entirely forensic in character. The team will at best establish that the apnoeic attack is caused by a violent assault. This is in fact a stated objective of the team. In some cases, the first recording of a suspected attack has produced less than satisfactory evidence, leading to further attempts. In one case the child was subjected to three such episodes before an evidentially trustworthy record was obtained. Forensic activity is usually and quite properly separated from the provision of care. The deliberate exposure of a child to attack in order to gather forensic evidence of this kind hardly ranks as good clinical practice.

The only other area of diagnosis which could justify the measure of impairing the capacity of the patient to determine what shall be done to her is the diagnosis of some form of mental health problem in the abusing mother. Such a procedure would be sanctioned by the Mental Health Act where the mother would be suspected of suffering from a mental disorder and thought to constitute a danger to the child. Under this section of the Act, the mother could be removed to a hospital and detained for assessment. This would not, of course, involve the child accompanying her, as the whole point of the exercise would be to protect the significant others, in this case the child, from the danger posed by the condition of the mother. It is not surprising, therefore, that the paediatric team does not invoke this provision for its diagnostic activities. This would frustrate the object of their technique. They argue that the patient is the child. Whilst it is understandable that their interest should be centred on the child, a psychiatrist would be concerned for the health of the mother. Munchausen syndrome by proxy now figures as a factitious disorder in DSM IV. Though the accounts of such cases claim that the abuser is not suffering from a mental illness, they assert that there is a personality disorder involved.[13] Setting aside difficulties we may have about such a diagnosis, of which more in a moment, it is worrying that the caring team in the cases in question neglects the possible interests of one patient altogether. This is especially significant when one takes into account the dire consequences of mistaken suspicions in alleged cases of child abuse by parents. Every care should be taken to protect the health and welfare of both parent and child in these situations.

When we consider the second arm of the supposed clinical intervention in covert video surveillance, we find that it too is flawed. The team's preferred justification for constructing impaired capacity in the mother is the protection of the health of the child, as part of the fundamental duty of the caring doctor.[14] We have seen that this takes the form of exposing the child to the expected life-threatening abuse, in order to provide evidence which will satisfy the courts that the mother and child should be separated, in order to protect the child from further attacks. There are other means of achieving this separation, as seen in the various court orders mentioned above. These do not involve the exposure of the child to violent attack, and though they may be more difficult to execute, they must be preferable in terms of protecting the health of both the child and the mother.

I have argued elsewhere[9] that the duty of the doctor to protect the health of patients is subject to limits determined by his or her professional role and expertise. These do not extend to protecting patients from violent assault. That protection is important and is the responsibility of others – in this case the police and social services, even though it may be the duty of the doctor to protect victims of assaults from the consequences of those assaults.

Can we find any support for the practice of covert video surveillance as outlined above, by comparing it with public health protection measures or Mental Health Act measures which enable the construction of impaired capacity to determine treatment, with a view to protecting others? Enough has already been said to suggest a negative answer to this question. Invoking the Mental Health Act measure of removing the mother to hospital, under an emergency order for assessment because of the danger she is thought to present to her child, would frustrate the objectives of the paediatric team. In any event, the category of disorder would be amongst those we have already questioned as being doubtful candidates for health care intervention – Munchausen syndrome by proxy being a personality disorder and not a mental illness.

The similarities with public health protection measures are superficial only, amounting to no more than the supposed element of preventive health care. We have questioned whether the prevention in the Munchausen case is of a health care kind, as it involves no special clinical expertise in detection or avoidance. It is designed to prevent injury, not disease or illness, and is centrally intent on establishing the commission of an assault. No disease condition is sought in the mother. Even if it was, and if Munchausen syndrome by proxy was accepted as a disease condition, it would not be that condition which would be transmitted to the child. Though there is some evidence that abused children become abusers, there is no reason to believe that this particular form of abuse is replicated in the abused children. Whilst the transmission of a disease is properly a doctor's business, *qua* doctor, the prevention of injury from violent assault is not.

CONCLUSION

We must therefore conclude that impairment of capacity to determine treatment decisions may be of two kinds. First, it may be rooted in a psychological deficit which undermines the ability of the patient to play an effective role in the decision-making process. Second, it may be constructed in order to protect others from the patient, who poses a danger to them. Not all of these constructions belong properly to the field of health care, as in some cases the threat and the means of its avoidance have no special relevance to the practice of medicine or its related disciplines. Health care professionals may find themselves pressed into the role of agents of social control through a blurring of the line between protecting health and preserving public order. This is most easily done in the area of mental health, where the important distinction between constructed impairment of capacity to determine treatment in the interests of others, on the one hand, and impaired capacity which is identifiable independently of the interest of others, on the other hand, is most readily scrambled.

University of Otago

NOTES

1 See Button, James T. H. (1994).
2 See Szasz, Thomas (1961).
3 See Sutherland J. D. (1973), p. 7.
4 See Walton, H. J. (1973), p. 26.
5 See Walton, H. J. (1973), pp. 30–31.
6 See Walton, H. J. (1973), p. 32.

[7] See Jones, Richard (1988), pp. 17–18.

[8] See Chatterjee, A. and M. C. Tosyali (1994), p. 1040.

[9] For a fuller discussion of this case see Evans, Donald (1995 and 1996).

[10] See *Family Court Reports* (1994), 2 FCR re. DH (minor) (care proceedings: emergency and order), 24th January.

[11] See Samuels, M. P., W. McClaughlin, R. R. Jacobson, and C. F. Poets, and D. P. Southall (1992), p. 165.

[12] See *Family Court Reports, op. cit.*

[13] See Meadow, Roy (1985), p. 392.

[14] See Southall, D. P. (1993), p. 614.

REFERENCES

Button, James T. H., *Communicable Disease Control: A Practical Guide to the Law for Health and Local Authorities*, Public Health Legal Information Unit, pp. 1–100, 1994.

Chatterjee, A. and M. C. Tosyali, "Munchausen's-by-Proxy," *The Lancet*, Vol. **343**, 1994.

Evans, Donald, "Munchausen Syndrome by Proxy and the Investigation of Life Threatening Child Abuse," *Journal of Medical Ethics*, Vol. **21**, No. 1, pp. 9–13, 1995.

Evans, Donald, "Covert Video Surveillance – A Response to Professor Southall and Dr. Samuels," *Journal of Medical Ethics*, Vol. **22**, No. 1, pp. 29–31, 1996.

Family Court Reports, 1994.

Jones, Richard, *Mental Health Act Manual*, London: Sweet and Maxwell, 1988.

Meadow, Roy, "Management of Munchausen Syndrome by Proxy," *Archives of Disease in Childhood*, **60**, pp. 385–393, 1985.

Samuels, M. P., W. McClaughlin, R. R. Jacobson, C. F. Poets, D. P. Southall, "Fourteen Cases of Upper Airway Obstruction," *Archives of Disease in Childhood*, **67**, pp. 162–170, 1991.

Southall, D. P., "Ethical Use of Covert Videoing for Potentially Life Threatening Child Abuse: A Response to Drs. Foreman and Farsides," *British Medical Journal*, Vol. **307**, 1993.

Sutherland, J. D., "Pathological Functioning of the Person: The Neurones and Psychosomatic Disorders," in: *Companion to Psychiatric Studies*, Vol. 2, ed. Alistair Forrest, London: Churchill Livingstone, 1973.

Szasz, Thomas, *The Myth of Mental Illness*, New York: Delta, 1961.

Walton, H. J., "Abnormal Personality," in: *Companion to Psychiatric Studies*, Vol. 2, ed. Alistair Forrest, London: Churchill Livingstone, 1973.

HANS-MARTIN SASS

ADVANCE DIRECTIVES: SELF-DETERMINATION, PHYSICIAN'S RESPONSIBILITY, VALUE OF LIFE*

RESPECT FOR PERSONS: HERITAGE OF THE SUBSIDIARITY PRINCIPLE

Controversies over values are nothing to worry about. Debates and discourses over systems of reference and arguments over those visions, values, and wishes which govern our individual life are the natural and essential expression of self-determination, self-responsibility, respect for others, and of the values and principles of liberty and security, as Spinoza would say; or of the values of liberty, justice, and the pursuit of happiness, as the fathers of the American revolution said. Discourse, understanding, controversy, and tolerance are the life and blood of peaceable and free societies, based on the respect for persons, the recognition of the individual conscience, and the risks, obligations and rights associated with choice and error. The alternative to our post-enlightenment and post-modern discourses and quarrels over values and visions would be hierarchical systems of heteronomous domination, torture, and ideational exploitation and tutelage. These systems, developed by self-proclaimed philosopher-kings and priest-czars for the exploitation of their fellow humans, beginning in pre-historic times, had been successful from the times of the pharaohs through the dark ages and middle ages of humankind, to the Hitlers, Stalins, and Pol Pots of our times.

It is only natural that value controversies and a rich diversity of visions and wishes in free societies would spill over into the fields of medicine, into the realms of the clinical and biomedical profession and into the existential challenges of each of us to deal with personal understanding and the handling of weakness, disease, pain and suffering, making babies, making others happy and being responsible for others, and coming to grips with our own terminality, death and dying. It would be very surprising – and very suspicious, indeed – if we all held the same convictions on issues such as recreative and "recreational" sex, the "value" of the different stages of foetal human life, the meaning of suffering, the criteria for death, and the attitudes towards prediction, prevention, and palliative care. Some societal and also

A.-T. Tymieniecka and Z. Zalewski (eds.), Analecta Husserliana LXIV, 239–253.

very personal value conflicts arise because many of us still feel obligated to search for watertight and binding orthodoxies and general solutions for problems which are "my individual and personal matter" and which have to be settled in accordance with my very personal visions and values, not by heteronomous ideational domination by professional organisations or systems of bureaucratic regulations or orientational dogmatics, creating additional conflicts between the individual's conscience and an overpowering outside world of heteronomous tutelage.

It was Spinoza who in 1670, in his *Tractatus Theologico-Politicus*, observed that peace and the fabric of society would not fall apart when individual freedom and liberty was granted, but rather on the contrary that peace, respect for persons and all other treasured values of a society rich in cultural and ethical values would fall apart if individual freedom was not granted. Tristram Engelhardt, in his *Foundations of Bioethics*, gives a wealth of arguments and information in support of Spinoza's enlightened vision. And for those who do not read books, the deeds of the McCarthys, Hitlers, and Karadzics of our times are evidence enough that the "people's power" recipe is superior to the "leader's power" model of yesteryear.

The prime principle for conflict resolution in value clashes between individuals and within free societies is "respect for persons" or, as the encyclical Quadrogesimo Anno (Pius XI, 1931, art. 79) puts it, the principle of *subsidiarity*, i.e. the right of the individual or the small and primary group to do good based on their individual conscience and calling. It requests that secondary groups or society in general honour individual conscience and withhold their own action and judgement when morally acting individuals accept their challenge. Josef Hoeffner has observed that the subsidiarity principle has two edges, one targeted at the limitation of ever growing bureaucracies and conceptual tutelage, and the other one at supporting and strengthening the individual's conscience and her self-responsibility and responsibility in the care of others (Hoeffner 1983). It is not necessary to quote the philosophical masters of the age of reason in support of the "respect for individual conscience" principle. Those of us who cling to traditional natural law models rather than to normative or utilitarian models of transcendental hermeneutics or discourse-and-contract models find evidence for that in reading the encyclical "Veritatis Splendor".

> Like the natural law itself and all practical knowledge, the judgment of conscience also has an imperative character: man must act in accordance with it. If man acts against this judgement or, in a case where he lacks certainty about the rightness and goodness of a determined act, he stands condemned by his own conscience, *the proximate norm of personal morality* (John Paul II, 1993, art. 60).

When transported from the field of social ethics into the realm of bioethics, this could be formulated in postmodern language in my first thesis: *Whenever politicians, theologians, philosophers or pressure groups fight over principles, theories and the preferred course of action, the primary moral agent, i.e. the persons closest to the moral challenge, should be given the right to follow their conscience and calling.* In the following, I will develop the rationale of introducing the principle of subsidiarity into one of the most crucial areas of conflicting values in contemporary clinical ethics: dealing with death and dying.

INDIVIDUAL CRITERIA VS. OBJECTIVE NORMS OF FUTILITY, DEATH, AND QUALITY OF LIFE

And here is my second thesis: *As long as philosophers, theologians, physicians and lay persons disagree on the bioethical interpretation of biomedical information on futility, criteria for death, quality of life, and the goal of medical intervention, the moral agent closest to the decision and most dependent on it has to be given priority in decision-making; in issues of death and dying, coma, and dementia, it is the person herself, not the physician, not the family, not society, nor the institutions of health care or health-care financing.*

Many controversies in academic bioethics and societal and cultural clashes over values are based in unsuccessful – and I should add: sometimes totally unnecessary and unwarranted – attempts to provide objective solutions for what basically are not objective, but personal and private, matters. This is particularly true for most moral challenges associated with the beginnings and ends of human life: issues of contraception, abortion, criteria for death, the meaning of suffering, and quality of life criteria. Of course, there are issues which warrant societal consensus and legal protection but their number is smaller than assumed by many. And where consensus in content cannot, but should, be achieved, consensus in methods for dealing with dissensus and protecting obligations and rights of individual value-based decision-making becomes the preferred means of protecting all three: individual conscience, trust-based provider-client interaction in the professional setting, and peace in society. What are the normative issues addressed by Advance Directives which are more appropriately resolved on a subjective rather than an objective level? I see the following: (1) criteria for the *salus aegroti* maxim; (2) risks and benefits of advance medical care planning and of giving directives in advance; (3) the authorisation of a trusted surrogate decision-maker; (4) resolving the issue of risks associated with definite directives for situations not yet fully understood or experienced.

(1) *Defining the Salus Aegroti.* Contemporary American bioethics, putting emphasis on the principle of patient *autonomy*, have identified the principle of self-determination as the prime reason for introducing model legislation in the support of advance directives. The "Patient Self-Determination Act" (1991) of the U.S. Congress intends to allow for the translation of the patient's autonomy into the processes of clinical decision-making by giving advance medical instructions and directives legal authority. As Olick puts it: "Since we are ordinarily the best judges of our own interests in terms of our own values and goals, extending patient autonomy also promotes patient well-being". But there is another line of argument, not primarily based on the patient's autonomy and more likely to be accepted by more traditional modes of reasoning in paternalistic physician's ethics of *benevolence.* This argument says that given the post-modern diversity in values and visions in a pluralistic society, a benevolent physician cannot define the "good for the patient" based on her own values, but rather needs to have information on the patient's values, visions and wishes in order to incorporate the patient's criteria for "good" into differential diagnosis, prognosis and intervention. As Sass et al. put it: "Advance care planning must be regarded as indispensable for good clinical practice, as it alone will provide definite necessary and vital information to define the salus aegroti, the 'good' for the patient". This leads to my third thesis: *The need to have advance directives for medical care is based not only on the principle of autonomy, but also on the principle of benevolence, as respect for patients as persons does not allow us to base diagnosis and intervention on medical-technical criteria alone, but requires the introduction of the patient's individual values, wishes and visions into individualized treatment decisions.*

(2) *Benefits and Risks of Advance Medical Directives.* There is some risk associated with making predictions in advance and prior to fully understanding the parameters and forces of future situations, in particular when the person has (a) no prior experience with those situations, (b) as a lay person does not fully understand decisions made in a highly technical and professional setting, and (c) future technical or other developments are beyond her control or not foreseeable. All three risks influence advance medical directives by lay persons: We rarely have prior personal and existential experience with situations of death and dying, dementia, severe suffering or coma. We don't understand the complex technical risks and uncertainties associated with medical procedures which we might refuse or request, and their consequences. We do not know the non-medical circumstances of situations which we have made distinct directives for, nor do we know about

the professional and technical expertise and options at that future time, as some of them might result from future clinical research or the lack thereof. In regard to advance health care directives, we know that medical experts, physicians and nurses, are quite reluctant to give very specific and inflexible directives which in a future situation might not be in their best interest and might actually be counterproductive to their "good".

(3) *Risks Associated with Designating Surrogate Decision-Makers.* These, again, are risks well-known in all areas of personal and professional life. It requires personal prudence, experience-based trust, competence and trustworthiness of the trustee, the exclusion of conflict of interest as much as possible, and some legal or other protection against abuse. These in themselves carry the risk of reducing the much needed authority of the proxy, or produce other additional forms of risk, conflict or uncertainty.

(4) *The "Old" and the "New" Person.* There is an extended debate as to whether or not previous directives should be honoured, or whether the "presumed actual will" of the patient at the time of intervention should guide intervention decisions. The normative conflict can best be illustrated by the German situation where a Chamber of Physicians guideline for the care of the dying requests that previous oral or written directives only be taken as a clue to determine the presumed actual will of the patient, while a Supreme Court decision holds that previous directives are binding and that so-called objective criteria of futility may only be used as a default position in the clear absence of the expression of individual preference [cf. Sass et al.]. A possible conflict between the previous and the actual person is real, but the controversy seems to be somewhat artificial and will have to be discussed in different scenarios: (a) As long as competent persons have not changed oral or written statements and directives, those statements should be taken as their true position and others should act accordingly in respect for persons; it is the right and obligation of competent persons who change their views and preferences to let others know, and if they do not do so they carry the risk of being misunderstood and mistreated. (b) If someone falls in a state of witlessness such as deep and prolonged coma, this person will have no new experience on which a change of values and wishes could be based and therefore should be treated according to wishes and values expressed. (c) Patients suffering from chronic and progressive illnesses and persons suddenly confronted with physical disabilities will or will not adapt to new and quite different parameters of quality of life. Clinicians are very aware

that many chronically ill patients and also those suddenly in a situation which they might have thought previously would not be worth living in, do indeed adjust to new challenges again and again. As long as these patients are competent, they have ample opportunity to accept or reject treatment; for those patients the use of advance directives is not indicated, and they, if they so choose, may adjust their previously stated preferences according to their new experiences and visions for life. (d) Severely demented patients, not knowing who they are, where they are, and unable to recognise friends and loved ones, who had executed advance directives while fully competent, refusing or requesting certain interventions in given situations, should be honoured as the persons they were when they made those decisions which they had then felt would be the most appropriate expression of their visions and values. (e) Hard cases, however, seem to be those where patients are semi-competent, where patients are in psychiatric confusion, or where incompetent patients request forms of comfort care which would contradict previous instructions. These are situations full of ambiguity in making the most appropriate benevolent clinical decision, and often the care for the "good" as presently expressed by the patient probably should be honoured over previous statements. Here is my fourth thesis: *Advance health care documents are the only valid expression of a patient's wishes, values, and visions as long as they have not been changed or invalidated.*

FORMS OF ADVANCE CARE DOCUMENTS AND DIRECTIVES

Given the normative challenges in preparing, executing, and reviewing advance care documents, there cannot be one single best model or form. Quite a variety of *advance directive* forms have been developed; they range from legalistic and rejectionist forms to value statements, questionnaires and checklists, and narrative models of value anamneses. A terse, comprehensive form has not yet been developed that could provide sufficient and adequate information and be flexible enough to be of advantage in case of various unforeseeable future events. Also, no form has been developed that persuasively solves the epistemological and prognostic problem of applying delicate values and wishes to unknown situations, a problem not unfamiliar in most fields of long-term planning and risk reduction. Basically, there are three different approaches: (1) the rejectionist legal approach, which forbids the physician to perform certain interventions in general or under certain conditions; (2) the proxy approach, or the designation of a trusted person to speak for the patient and to request or reject certain procedures when the time has come and the situation is less uncertain; (3) the value history approach, in which the patient provides information about his or her values and wishes that can later be included in the medical history and can thus provide the necessary differential diagnosis on which prognosis and intervention decisions can be based for individualised treatment. Finally there are (4) a combination of different forms and variations on the above approaches. All models have their advantages and disadvantages.

Rejectionist Legal Directives: This type of advance directive, characterised as a legalistic and rejectionist document, widely distributed in the U.S. and using language provided by State law, is difficult to translate into clinical reality. Nevertheless, the documents are useful in that they give loved ones and care-givers a sense of direction. For instance, such a document might indicate that a patient does not want intensive, severely sedative, painful prolongation of life, even if the exact circumstances are either not clearly defined (if vague terminology such "no extraordinary means" is used) or too narrowly fixed (if extended lists of rejected procedures for specific situations are presented). Physicians quite often are reluctant to follow or even to recognise detailed rejectionist requests, as there is suspicion that they might not apply to the specific situation and therefore might not correlate with the presumed actual will of the patient. It is my (fifth) thesis, that *rejectionist living wills written in legal language are difficult to translate into clinical practice and therefore might not in be in the patient's best interest.*

Designation of a Proxy Decision-Maker: The proxy form is well known and well accepted. In most models, it is typically accompanied by some rejectionist statements. But by itself, a designation of a surrogate decision-maker is a valid and flexible instrument that prevents the harm that may be done by those overly specific living wills that are often drafted long before the specific situation arises. However, if the proxy fails to understand and to represent the patient's values adequately, it can be one of the more risky instruments. When used in combination with a durable power of attorney for health care, rejectionist documents can become more flexible and easily adaptable to specific situations. It should, however, be clearly articulated in the written document which voice shall have precedence in the event that conflict arises between the physician or the care team, the surrogate or the written document.

Belief and Value Status Forms: The value history document uses checklists of general values, wishes and preferences, in particular those associated with medical care for the terminally ill or the comatose. A Value History (Wertanamnese) is an instrument that establishes a person's value profile, which can then (a) be used by the patient to establish a more explicit directive, (b) be attached to a formal living will as an instrument for its interpretation, and (c) be used by the surrogate decision-maker as a guidance tool. Most checklists or questionnaires have been developed in conjunction with formal advance care instruments. These questionnaires can address suffering and end of life issues, thereby helping to overcome the denial of a topic that is considered taboo; these are good instruments with which to generate discussions within the family and between the patient and the prospective surrogate or the health care giver. Questionnaires do not replace formal advance directives, as they do not address conditions in which the value statements and treatment preferences will be used.

Here are two lists that address general and specific values and wishes:

(A) *Basic Life Values.* Perhaps the most basic values in this context concern length of life versus quality of life: Which of the following two statements is the most important to you? 1. I want to live as long as possible, regardless of the quality of life that I experience. 2. I want to preserve a good quality of life, even if this means that I may not live as long.

(B) *Quality of Life Values.* Many values help to define the quality of the life that we want to live. The following list contains some that appear to be the most common. Review this list (and feel free to either elaborate on it or add to it) and circle those values that are important to your definition of quality of life: 1. I want to maintain my capacity to think clearly. 2. I want to feel safe and secure. 3. I want to avoid unnecessary pain and suffering. 4. I want to be treated with respect. 5. I want to be treated with dignity when I can no longer speak for myself. 6. I do not want to be an unnecessary burden on my family. 7. I want to be able to make my own decisions. 8. I want to experience a comfortable dying process. 9. I want to be with my loved ones before I die. 10. I want to leave my loved ones with good memories of me. 11. I want to be treated in accord with my religious beliefs and traditions. 12. I want respect shown for my body after I die. 13. I want to help others by making a contribution to medical education and research. 14. Other values or clarification of values listed above ... (Doukas, McCullough). This is my sixth thesis: *Physicians and nurses do not need to be told by lay persons what to do in a specific future situation, but they need to know about their patient's visions, values, wishes and concerns, otherwise they cannot provide for individualised personal treatment and the patient's good.*

Combination Forms: Probably the *best method* for identifying, promoting and protecting one's own values and wishes is to invest time in the process of communicating with loved ones, trusted physicians and friends, and drafting a *combination form* that is not too inflexible but provides adequately precise insight into one's own value reasoning. These forms should then be validated with a physician, routinely reassessed, and eventually readjusted based on additional experiences and changes in value priorities. Just as a Last Will might improve when devised and reviewed by a legal expert, so does the Living Will require consultation and review by a medical expert. An informative, directive and still flexible Living Will has to contain four segments in its anatomical structure: (1) information on personal values, hopes and angst; (2) a small list of medical directives or instructions; (3) designation of one or more proxies; (4) validation issues.

(1) The most important information for clinical decision-making is information on the potential patient's value-and-wish status. *Values* and wishes can be stated in religious or secular terminology. Values can be stated in different ways, narratively, by checklists and by description. The more informative the document is, the better it can be used in clinical hermeneutics to introduce bioethical information together with biomedical information into differential diagnosis and prognosis, thus determining individualised and patient-oriented treatment decisions and review.

(2) As far as *instructional medical directives* are concerned, it is advisable to keep instructions brief and clear, and allow for flexibility. *Scenarios* to be addressed are: (a) preterminal and terminal stage with irreversible and often painful underlying diseases and various forms of multimorbidity, (b) prolonged coma and persistent vegetative state, (c) severe dementia. *Issues* to be addressed include: (a) palliative care, (b) withholding or withdrawing of interventions, (c) acceptance of the "double-effect" of medication which unintentionally might allow for or hasten death, (d) withdrawing of basic support such as ventilation, nutrition, and hydration, (e) wishes regarding the social and living environment. The administering of best palliative care, the double effect which certain medications might have, and the respect for patient's wishes regarding the environment he prefers to live and to die in are fully accepted by a wide majority of religious and secular bioethicists and physicians. Withholding of technically possible intervention at the request of the patient or her proxy is easier to accept than the active termination or withdrawal of an already initiated intervention, as there is a large emotional and a minor ethical difference between passive "acceptance" (Unterlassen) and active "doing" (Tun). The clinical admin-

istration of nutrition and hydration can either be understood as medical intervention or as basic care and an act of basic human solidarity; the latter position will have ethical problems in withholding medically administered life support, in particular hydration. A recent clinical study, however, reversed the traditional understanding that dehydration at the end of life would be associated with additional painful suffering; the study shows that dehydration results in a euphoric rather than a painful state; these findings might increase the number of clinicians and bioethicists who would honour clearly expressed wishes to even withhold hydration and to allow for a naturally occurring death without medical intervention. It is self-understanding that even in situations of withholding basic life support such as hydration, comfort care such as the moistening of lips and mouth has to continue and be given fullest attention, as comfort care will be the last and only avenue left for care, support, and compassion. As future medical scenarios are impossible to predict, it is not reasonable to be too detailed in giving medical instructions for yet-to-be-determined and described clinical situations. But patients expecting predictable developments of pre-existing conditions might prefer to address those situations, thus sparing loved ones from making hard choices later on; e.g., patients suffering from amyotrophic lateral sclerosis might express the wish not to start ventilation support ever. Also, there are atypical treatment preferences or rejections, such as those based on religious beliefs, that should be mentioned definitely and the reasons be given clearly and without hesitation. But the longer the lists of rejections of medical interventions are, the more likely it is that they might be counterproductive and not in the reasonable interest of the patient, as the risk of inflexibility is too high.

(3) *Surrogates* are much more flexible to act in the patient's interest. Surrogates are of particular importance when patients become demented. This draft does not present medical directives or options for prolonged coma, dementia, and severe and irreversible results of severe strokes, as these issues are probably best discussed in partnership-in-trust between the physician and the surrogate when interpreting the patient's value profile. But, of course, medical instructions for comfort care can only be given for the situations of persistent vegetative state or for full blown dementia ["when I do not know who I am and where I am and do not recognise friends and loved ones any more"], as my own Living Will does.

(4) Finally, Living Wills should not just be witnessed, but, the more detailed they are, also *validated* by reference to and a review by a trusted physician who has explained terminology, options, consequences of refusal

and who witnesses in regard to the sound state of mind of the person executing the Living Will.

And here is my last (seventh) thesis: *Advance directives are indispensable for providing the best possible care for those who cannot express their treatment preferences anymore. Advance directives should inform the health care professional and other surrogate decision-makers primarily about values and wishes which shall govern treatment, nominate proxy decision-makers, and refrain from detailed medical instructions, except in the rarest cases where existing and progressing disorders allow for definite instructions for the future course of continuing or withdrawing treatment.*

INSTEAD OF A SUMMARY: DRAFT OF A CHRISTIAN LIVING WILL

Living will forms, written in secular or in religious language, typically express the person's values and wishes, designate proxies, and direct, from a lay person's perspective and understanding, the preferred medical interventions in a state of incompetence or in the preterminal or terminal phase of life. They should be informative and instructive enough to provide guidance, but they should not be too inflexible to allow for compassionate clinical decision-making for the good of the patient as expressed in the patient's living will.

Instead of a more theoretical assessment of the anatomy of the most well-designed living wills and the challenges involved in their execution, let me present a draft of a *Value Statement and Medical Directive*, written in religious language and submitted for discussion at this conference. In drafting this document, I have used a form proposed by the Evangelical-Lutheran churches in Bavaria and Thuringia and printed in the official Hymn Book of these two churches, a form developed by the Committee on Jewish Law and Standards in New York called "Instructions for the Valley of the Shadow", and the original form designed and distributed by the Choice in Dying group in New York since 1979 (all in: Sass, Veatch, Kimura, 1997). The draft that I am proposing is intended to be brief, informative, and flexible for medical interpretation within the normative limits set out by the signee. The form consists of four parts: (1) a general expression of beliefs and values, which make the person known to those who will treat her later, (2) a proxy declaration designating a person and delineating this person's authority in future medical decision-making, (3) a brief list of medical directives. It is advisable to have the form witnessed according to laws and regulations. If specific medical directives are given, such as for expected situations of existing chronic failure, for resuscitation, or for the withholding or withdrawing of nutrition or hydration, those should be discussed with a trusted physician beforehand. It is also advisable that a health care expert certify that the signee is in a state of mental health and understands the clinical consequences of her directives.

MY VALUES AND WISHES WHICH SHALL GUIDE MEDICAL TREATMENT

Part One: My Beliefs and My Values

"The LORD gave and the LORD has taken away; blessed be the name of the LORD". I believe that my time and my life are in GOD's hands; I live and I die by GOD's will. If a future situation arises where I might not be able to communicate my wishes, values, and visions to family, friends, physicians, and those who will care for me, I want the following directives to be understood as my final and last expression of wishes and values. Those who will have to care for me medically and humanly must not do everything possible to keep me alive, nor should they treat me in the way they would treat others or would be treated themselves, but instead should base their decisions on the following information.

Part Two: Directives for Medical Treatment

In order to protect my life, a gift from GOD, it is my wish that all prudent medical treatment be provided to me with the aim of effecting my recovery or reducing the effects of chronic illnesses or disorders. Should that be deemed impossible, I request compassionate nursing care and the alleviation of suffering to the extent of all appropriate measures only. If interventions necessary to provide comfort care and the alleviation of suffering cause harm to my body or shorten my life, then let it be, as this may be God's way to take from me what HE has given. Under no circumstances, however, do I want to be intentionally killed, nor do I ask that my life, a gift from GOD, be taken on the basis of my presumed consent or request. Should I be in unrelenting and unrelievable suffering or pain, then I request that I be allowed to die and let the LORD take what HE has given. If such a situation occurs, I request all possible comfort care and the medical alleviation of suffering associated with my condition as death approaches. If I fall into prolonged coma or if physicians diagnose that my life will continue in what is called a persistent vegetative state only, than I request comfort care only. If I become mentally

incapacitated so that I do not know who I am and where I am and do not recognise loved ones and friends any more, then I want/do not want my life to be prolonged by extraordinary means such as major diagnostic or therapeutic interventions. As I know that my life is in GOD's hands, I request that "extraordinary means" such as technical ventilation and artificial nutrition and hydration be withheld or withdrawn in situations of unrelenting and unrelievable pain (yes/no), prolonged coma or persistent vegetative state (yes/no), or the severest forms of dementia (yes/no). I ask for your prayers and for religious support in all situations of weakness, incompetence, suffering, and dying.
If possible, I would like to die in surroundings familiar to me. Regarding pre-existing medical conditions, I request/suggest the following: ...

Part Three: Designating a Surrogate

If in a future situation I am no longer capable of expressing my values and wishes, I designate ... (name and full address) to be my legal and health care representative. I recommend/request that my representative seek and follow the advice of ... (name) in religious and value matters, of ... (name) in medical matters, of ... (name) in legal or financial matters. If a conflict arises between my representative, those other trusted persons I have named and/or the medical team or the physician with primary responsibility, or if their interpretation of my values and wishes differ, I hereby declare that the final authority shall rest with my representative/Physician/other: ...

Part Four: Validation and Signature

This declaration, which expresses my beliefs, values and wishes, shall guide others when I am no longer competent to decide for myself. I have discussed specific medical directives with a medical expert (you may want to give the name ...) and understand the choices I have made. I intend to review and to re-validate this declaration from time to time, but as long as I have not modified or nullified it, it shall be legally binding as the final expression of my values and requests. Copies of this declaration are available at: ... I express gratitude to all who care for me with expertise, compassion, and prayers during the last weeks and days of my life.

... (signature), ... (address)

Not everyone believes in God as Christians, Moslems, or Jews do. But for everyone there is the real fear that medical experts do too little in certain situations or too much in others, based on objective standards of medical futility that are not in accordance with one's own visions, values and wishes for the last hours of one's life. The issues to be addressed by one's fellow humans, regardless of whether their orientation is religious or secular, are more or less the same as far as they concern advance medical directives. Instead of affirming the belief that one's life "is in God's hands", a non-religious person might express the values upon which her rejection of life-prolonging intervention "by all means" is based with the words "I know that human life is terminal and must come to an end sooner or later". Those of us who believe that life is a gift from God that must be taken care of and not terminated of one's own free will, do not accept physician-assisted suicide; but there are other systems of reference which definitely will agree that self-determination does not exclude the right to freely end one's own life if this would be more adequate to the person's visions and values for life than just the sheer prolongation of life under conditions not acceptable. It will be a different task, however, to find a health care expert who, in her understandings of benevolence, will assist in "self-delivery", or "put the patient to rest" at the patient's request.

For all mortal human beings, whatever their system of value reference, the words of the Psalmist cannot be disputed: "The years of our life are threescore and ten, or even by reason of strength fourscore; yet their span is but toil and trouble; they are soon gone and we fly away", (Ps. 90). These words remind us all that we are masters in many fields, including the field of medicine, but that we are not masters of death and that it would be arrogant to fight aging or dying as we do disease, rather than accepting both as a natural fact or a divine law.

*This study was supported in part by a grant from the Volkswagen Foundation to the Institute of Philosophy, Ruhr University, Bochum, and to the Kennedy Institute of Ethics, Georgetown University, Washington, D.C.

Ruhr University, Bochum

BIBLIOGRAPHY

Bundesarztekammer 1993. "Richtlinien für die ärztliche Sterbebegleitung". *Deutsches Arzteblatt* 90 (37): C1628–9.

Bundesgerichtshof 1994. Urteil vom 13. September 1994. *BGH – 1 StR 357/94 – LG Kempten*, Karlsruhe: BGH.

Cantor, N. L. 1993. *Advance Directives and the Pursuit of Death with Dignity*, Bloomington: Indiana U. Press.

Doukas, D. J., McCullough, L. B. 1991. "The Values History. The Evaluation of the Patient's Values and Advance Directives". *Journal of Family Practice* **32** (2): 145–153.

Doukas, D. J., Reich, W. 1993. *Planning for Uncertainty. A Guide to Living Wills and Other Advance Directives for Health Care*, Baltimore: John Hopkins U. Press.

Engelhardt, H. T. 1996. *The Foundations of Bioethics*, New York: Oxford U. Press, 2nd ed.

Eser, A., Koch, H. G. 1991. *Materialien zur Sterbehilfe*, Freiburg: Max Planck Institut für ausländisches und internationales Strafrecht.

John Paul II, 1993. *Veritatis Splendor*, Vatican City.

Hoeffner, J. 1983. *Christliche Gesellschaftslehre*, Kevelaer.

Kielstein, R., Sass, H. M. 1993. "Using Stories to Assess Values and Establish Medical Directives". *Kennedy Institute of Ethics Journal*, 3: 303–325.

King, N. M. P. 1991. *Making Sense of Advance Directives*, Dordrecht: Kluwer.

Koch, H. G., Meran, J. G., Sass, H. M. 1994. *Patientenverfügung und stellvertretende Entscheidung in rechtlicher, medizinischer und ethischer Sicht*, Bochum: Zentrum fur Medizinische Ethik.

Olick, Robert S. 1991. "Approximating Informed Consent and Fostering Communication: The Anatomy of an Advance Directive". *Journal of Clinical Ethics* **2** (3): 181–189.

Pearlman, R. A., Cain, K. C., Patrick, D. L. et al. 1993. "Insights Pertaining to Patients' Assessment of States Worse than Death". *Journal of Clinical Ethics* **4** (1): 33–41.

Pius X. I. 1931. *Encyclical Quadrogesimo Anno*, Rome.

Sass, H. M. 1994. *Hippokratisches Ethos und nachhippokratische Ethik*, Bochum: Zentrum für Medizinische Ethik.

Sass, H. M. 1998. "Images of Self-determination and Beneficence, of Killing and Letting Die". *Advance Directives and Surrogate Decision-Making in Multicultural Perspective*, Sass, H. M., R. M. Veatch, R. Kimura, eds., Baltimore: Johns Hopkins University Press.

Sass, H. M., Bonkovsky, F. O., Akabayashi, A., Kielstein, R., Olick, R. S. 1996. "Advance Health Documents in Multicultural Perspectives". *Jahrbuch für Recht und Ethik / Annual Review of Law and Ethics*, in print.

SUPPORT Investigators. 1995. "A Controlled Clinical Trial to Improve Care for the Seriously Ill Hospitalized Patients". *JAMA* 274: 1591–1598.

Teno, J. M., Nelson, H. L., Lynn, J. 1994. "Advance Care Planning: Priorities for Ethical and Empirical Research". *The Hastings Center Report* **24** (6): S32–S36.

U.S. State Court of Appeals for the Ninth Circuit. 1996. "Compassion in Dying vs. State of Washington", *US State Court of Appeals. No 94-35534. D.C. No. CV-94-119-BJR*, Opinions 3109ff, 3161ff, 3219ff.

BARBARA MAROSZYŃSKA-JEŻOWSKA

THE ETHICAL PROBLEMS OF DOCTORS' DECISIONS: THE PATIENT'S AUTONOMY AND PATERNALISM

Modern medicine has forced physicians to face numerous problems unknown to previous generations of doctors: the participation of patients in drug testing, informed consent, the question of the physician's truthfulness in informing patients about their health condition, or the problem of the active participation of patients in the process of making decisions concerning their therapy.

Clinical decisions made by physicians in their everyday work have a definite ethical aspect, as they are not limited to purely scientific problems like diagnosis or a selection of the most effective therapy. Medical ethics is situated on the borderline between medicine and philosophy. It is one of those areas in which cooperation between physicians and philosophers is necessary. In this paper, selected medical ethics issues are discussed from the point of view of philosophy, i.e. certain philosophical notions or conceptions are applied in the interpretation of specific medical problems. Experimental sciences – and biological medicine is one of them – are always based on certain theoretical assumptions, and are never free of certain value-based convictions. Much is being done to prove that experiments are free of subjective factors, but the final interpretation of the quantitative results of such research must always take into account the researcher's earlier convictions. The probability of the truth or falsehood of a scientific hypothesis is always subjective.[1] This means that when we decide which hypothesis to confirm and which one to reject, we can never eliminate the subjective factor.

Medicine is not only a science, it is also a practical discipline. A physician may be deeply convinced that he has objectively presented (to his/her patient) a description of the results of medical examinations conducted, and a diagnosis. Yet, it is possible that the patient, treating his illness only to a limited extent as a mechanical injury to his organism, does not think that everything is clear as soon as the problem is diagnosed. S/he wants to know about those elements of everyday life which have caused the illness. Usually physicians only examine the functioning of the human body. They completely neglect the relationship between function and consciousness. I accept the view that illness is not merely a biological concept where illness is seen as a defect in the functioning of a "biological machine", a deviation

A.-T. Tymieniecka and Z. Zalewski (eds.), Analecta Husserliana LXIV, 255–262.

from a model, pattern or norm specific for the given species,[2] but that it includes other elements.

People seeking a doctor's advice indicate a subjective feeling of illness or good health. The mechanistic model, an important and necessary way of viewing illness, does not adequately explain the concept of illness. Although illnesses are biological defects, their causes, symptoms and effects reach far beyond the boundaries of biology. In clinical practice, a physician must take into account the pain experienced by the patient, his/her suffering, his/her sense of dignity, and his/her understanding of the meaning and quality of life. The physician must also learn to treat such extra-biological phenomena in a natural way. This is also related to a definite conception of a human being that is not limited to biological and behavioral terms (although this naturalistic conception has exerted an essential influence on numerous fields of science).

A physician faces human suffering, fear, despair, i.e. phenomena belonging to the subjective sphere. S/he must recognize the patient's autonomy, if s/he considers that a person is a thinking individual aware of his/her identity. Obviously, I do not deny that a person is a biological organism. I only wish to stress that if we take into account hermeneutic thought, then a person is not merely a biological organism, and medicine is something more than a branch of the natural sciences.[3] One can strive to determine a person's constituent components and state that s/he is an individual capable of free action. This goes further than stating that s/he is a specifically formed biological or social being. A proper understanding is possible only through philosophical reflection. We should note that neither Kierkegaard nor Heidegger reject the necessity of examining a person in an empirical way, as is done in modern psychiatry or medicine. A proper interpretation of the empirical data requires a reference to hermeneutic reflection (the phenomena of fear or depression can serve to illustrate all the limitations of the naturalistic approach).

Kierkegaard claims that a person is a synthesis of body and soul – a synthesis of the spiritual element and the material one. This part of his theory does not differ much from the conception of a person as a biological and social individual. It is also in accordance with the holistic theory of modern psychiatry stating that an illness is the result of the action of numerous interrelated biological and psycho-social factors. Thus, the notion of synthesis can be interpreted as a combination of many different factors, and in this respect this theory is related to modern naturalism.

A person, however, is something more than just such a synthesis, for there exists a third element which unites and completes the synthesis, which makes a person a conscious being – it is either the Spirit or the Self. The Self or, as it is now called, personality, is this third element which makes up a person and determines his/her characteristics, and his/her unique form of existence. Hence a person has a sense of his/her own identity, and thanks to this he/she perceives him/herself as an autonomous and free individual. Thus, self-awareness and awareness of freedom are the essence of this third, integrating factor. A person is a self-aware being who is capable of reflection (self-reflection) on his own feelings, desires, and who can fully realize how they correspond with the values s/he has chosen and esteemed, and with the conditions in which s/he lives.

The importance of subjective truth is recognized in modern medicine. Objective truth or objective reflection is of little importance when considering an individual's existence. Although a person is frequently thought of in a naturalistic way, s/he cannot be fully understood in this way. Yet many physicians are inclined to attempt to do so. Features such as will, freedom, understanding, and fear cannot be adequately explained if a person's nature is seen in purely physical-somatic terms. A physician must take into consideration not only anatomical and physiological defects, but must also take into account the importance which the patient attaches to his illness and its symptoms in the context of his/her whole life. We expect the moral attitudes adopted by a person – by a physician too – to be in accord with his/her meta-ethical principles. The person who deals with medical ethics must also take normative ethics and meta-ethics into consideration. Meta-ethical assumptions may be different: they may state that morality is a product of human beings or of social relations, or that it is given by an external criterion – and the moral system is something which we are just discovering.

Two basic types of normative ethical theories exist: utilitarian and deontological ones. Some claim that it is only the results, the consequences of our actions that count (utilitarianism). Others try to prove that we are morally obliged to respect certain principles (e.g., certain laws and duties) regardless of the consequences. This is known as deontological ethics. Numerous problems encountered in medicine are just conflicts between the utilitarian and deontological points of view. One of the most difficult problems of moral philosophy is deciding whether these two standpoints can be reconciled. The physician who faces an ethical problem must decide what s/he should do and what is best for his/her patient. It is easy here to adopt a paternalistic attitude toward the patient. Its justification may be either utilitarian or deontological, depending on what interpretation of the term "good" has been adopted.

Naturalistic meta-ethics[4] is closely related to utilitarian normative ethics. The application of the theory of usefulness to medical decisions is usually limited to an analysis of profits and losses. It is considered obvious that a physician makes decisions on behalf of the patient, in accordance with the principle of utilitarianism orientated towards the patient's good. The final choice of therapy is based on the physician's overall assessment of the influence of all possible factors, i.e. both favourable and unfavourable, on the patient's quality of life. The second aspect of the physician-patient relationship, namely the patient's autonomy, has been neglected here. If we take an extreme view of the patient's autonomy, we may say that the patient is the only person who can fully evaluate all the positive and negative aspects of possible therapies. As an autonomous person, he has the right to choose that which he considers most beneficial for him/herself. In this context we recognize the physician's duty to respect the patient's rights or, at least, not to neglect the rights of a patient as an autonomous person (independent of the physician's medical knowledge). The above remarks indicate the sources of utilitarian and deontological thinking. The basis of this conflict is the necessity for the physician to take into account the consequences of his actions while simultaneously paying attention to the patient's subjectivity. The concept of "autonomy" is ambiguous; it may have different meanings, depending on who uses the word. Kant and J. S. Mill are often put forward as people with opposing interpretations of the word.

For Kant, autonomy is a constituent property of the human being, and is closely connected with the concept of the individual's freedom. Every violation of a person's autonomy is equivalent to the violation of humanity and treating this person as a thing. J. S. Mill, one of the founders of utilitarianism, pays more attention to the issue of an individual's freedom. Meanwhile, the person who accepts Kant's radical standpoint is in opposition to the utilitarian view as regards justification of the principle of paternalism. It can be summarized as permission to act on another person's behalf when we are convinced that it serves his/her interests (benefits) best. What strikes us is the indeterminacy of the concept of benefit in the context of the impossibility of its neutral formulation. This does not mean, as the authors of *Philosophy of Medicine* suggest,[5] that paternalism, in the broadest sense of this word, is or should be rejected. They indicate three types of paternalism: true paternalism, justified paternalism and unjustified paternalism.[6] Natural justification of paternalistic behaviour (this relates to true paternalism) includes situations of highly diminished autonomy of the patient, e.g. mental retardation, blurring of consciousness due to high temperature, the patient's loss of consciousness, etc. Justified paternalism is based on the conception of the patient's clearly expressed or presumptive consent. The patient has a right to put himself in the physician's hands and act in accordance with the doctor's instructions. Therefore, he can expect the physician to do everything that is needed on the basis of his professional knowledge. It is unjustified paternalism which is most controversial from the moral point of view. The basis for such actions on the part of physicians is usually their personal view of a person, their intuitive system of values, and their medical knowledge and intuition. In this way the physician's attitude reduces the patient's ability to use his/her autonomy. What makes things even more difficult is the lack of one socially prevailing morality. Nowadays a physician cannot assume that the patient's attitude towards life, illness and death is similar to his/her own, and vice-versa.

One general difficulty is distinguishing between justified and unjustified paternalism. People usually stress the need for a public debate and empirical ethical investigations, which would show what patients really expect. These investigations could help utilitarian-minded physicians to broaden their understanding of the terms "benefit" or "the patient's interest". The necessity of studying the structure of social morality, and hence aiding the selection of priorities, is also emphasized. We must reject the claim that every illness is a state of limited autonomy, which directly justifies the paternalistic attitude. A specific example of the patient exercising his/her personal rights is his/her active participation in all phases of the decision-making process. This means free informed consent to therapy, medical experiments or controlled clinical tests. It is a particularly delicate issue to determine what is and what is not mature consent based on reliable and continually updated information. We should also take into account the human inclination to make use of other people, albeit unconsciously, with possible future benefits in mind. Taking this into consideration, we must agree that achieving a reasonable consensus is the minimum necessary in the relationship between a physician and a patient. It reflects the patient's need to participate in decisions concerning his/her own treatment. In cases where consensus cannot be achieved at the start of therapy, consent to treatment should be acquired later, when the patient is finally able to express it. A special issue in clinical examinations is the so-called method of controlled clinical testing, when the utilitarian approach (treating the patient's autonomy as a certain good that can be "sacrificed" for a greater good, e.g. better results achieved with the future patients) is replaced by the radical deontological approach giving the patient full autonomy (i.e., the physician should obtain voluntary and conscious consent based on reliable information). Sometimes a physician, not without some justification, uses his/her professional position, and informs the patient that s/he is unable to understand the complicated – from the point of view of medicine – stage of the illness, and s/he must be satisfied with the information that everything is being done to help him/her. In the past it was assumed that there was a common, established morality. A doctor could be firmly convinced that his/her patient's convictions as regards suffering, death, and human dignity were basically the same as his/her own. In turn, the patient could justly expect that his/her convictions would be respected. Therefore, the patient, in submitting to medical treatment, could feel that by doing so he would not be giving up his moral autonomy. In a situation where the liberal pluralistic morality dominates, this relationship is not so unequivocal. Besides, assessment of the value (quality) of individual life does not lie within the scope of medicine. In this context we encounter the problem of unnecessary suffering and prolongation of this suffering by doctors – the controversial problem of "orthothanasia". Thus, in medical practice there exists a departure from the principle of unquestionable prolongation of life, when physicians consciously (not due to negligence) stop administering life-maintaining drugs to a patient in agony. The physician chooses this moment him/herself, on the basis of his/her professional knowledge and his/her convictions. Some societies are being formed, whose members sign declarations asking that in the case of their developing an incurable illness they should not have their lives prolonged by technological or pharmacological aids, and that physicians limit their actions to relieving the suffering (and allow a death free of suffering). The controversial problem of unnecessary suffering exists, and, paradoxically, it is also a reflection of the physician-patient relationship in the context of paternalism vs. autonomy.

In a "traditional" physician-patient relationship, the decision is made by the physician for the patient. In a partner-partner relationship the decisions concerning treatment are the patient's. Thus, s/he must be clearly and exhaustively informed about the illness and all possible ways of treating it. Yet, even this will not end the problem of the patient's dependence on the physician completely, because it is not the patient who defines his/her needs and expectations; this is done for him/her by the physician who provides the medical service. Partnership in the patient-physician relationship started to develop as early as the 19th century. It was forced into existence by patients disappointed by particular physicians, the possibilities of medicine, or errors in medicine.

Everything suggests that the paternalistic relationship will start to disappear. Rapid changes in morality are taking place, respect for various authorities is disappearing, the distance between patient and physician is decreasing. The patient's right to information about his/her health and his/her right to make decisions concerning his/her therapy are gradually being included in legal acts and physicians' codes of ethics. Various forms of patients' organizations are coming into being (self-support groups, patients' associations, national federations of such associations).

It is hard to convince a physician to like the idea of partnership, and not only because it is easier for him/her to be the patient's master. It was Paracelsus who indicated that a physician is fully a physician when s/he becomes a patient. Authority and omniscience are instruments, thanks to which the therapy seems to be more effective. Health protection, saving life, and relief of suffering are goods which a physician is obliged to provide to each of his/her patients without discrimination. Traditionally it was the physician's moral obligation to provide these services, rather than the patient's right to have them. Today, the situation is changing – the view is spreading that every (wo)man is entitled to demand medical care and to influence its course. Nowadays, the right to receive definite goods and to make unconstrained use of them, granted to every person, is starting to reach far beyond the traditional sphere of medical activity.

Unfortunately, observations in the world of medicine show that the changes in the system of values, the imperfection of philosophical ideas and their consequences in life, are greatly weakening (wo)man's personal identity. We are still a long way from providing holistic treatment for people and respecting patients' needs.

On the one hand, a physician must become part of a technological apparatus in order to achieve effectiveness in diagnosing and curing numerous illnesses. On the other hand, however, (s)he is obliged to recognize the patient as a person possessing the inalienable right to enjoy his/her subjectivity.

The "ideal" attitude, combining the best possible treatment with respect for the patient's dignity, freedom and feelings, has its roots in human fraternity. Curing an illness is a scientific and professional activity, while treating a patient should involve the influence of the physician's personality on the patient, i.e., what Nielubowicz calls care for a patient. In such a physician-patient relationship there is respect for the patient's subjectivity. In order to learn about ethical problems in medicine, the modern physician needs access to some integrated system embracing many disciplines, both philosophical and biological/medical. Modern medicine should overcome its reductionist and neo-positivist limitations, take a critical look at itself, and free itself of various dependencies, which means to gain self-consciousness. A physician – despite the observed tendencies – should be not only a pragmatic professional, but also a person consciously and devotedly serving the patient's health. In this broader perspective it will be easy to find room for a more comprehensive understanding of sick people, by treating them as autonomous subjects, with all the consequences of this fact.

Kraków

NOTES

1 In everyday language, probability is our subjective conviction regarding the occurrence of a given event; it is different from statistical probability.

2 A. Ross Sygdamsbegrebet, *Bibliothek for Laeger*, 1979, No. 171, pp. 111–129.

3 H. Wulff, S. Perdersen, R. Rosenberg, *Filozofia medycyny*, Warszawa: PWN, 1995, p. 160.

4 Supporters of this view believe that words like "good" and "should" can be reduced to the properties of the world of nature, like "happiness" and "desire".

5 H. Wulff, S. Pedersen, R. Rosenberg, *Filozofia medycyny*, op. cit., p. 237.

6 Ibid.

JACEK JAŚTAL

SOME REMARKS ON "QUASI-SURROGATE" DECISION-MAKING

Taking an active part in patients' decision-making is inevitable in the medical profession. The role played by the doctor can range from a "paternalistic" one to a "subordinate" one: paternalistic decision-making is oriented towards the patient's well-being (disregarding the patient's own opinion), and subordinate follows the patient's choices, even if, in the doctor's opinion, these are harmful, or not as conducive as other alternatives would be. Both stances have rather obvious drawbacks: the first leads to restriction of the freedom of the patient, the second may lead to "objective" harm being done to the patient. Between these two extremes lies a whole spectrum of solutions. Because of the difficulty in reconciling the two above-mentioned factors it is almost impossible to work out a practical set of guidelines for doctors (which would not be subject to criticism either for ignoring the patient's will, or his good).

There are cases in which the patient is unable, at least for some time, to take part in decision-making, however important for him the decisions might be. Extreme examples include infants, persons with advanced Alzheimer's disease, those suffering from mental illness, or those who are mentally disabled or in a coma. Then, the burden of decision-making must be borne by others: the patient's relatives, people legally responsible for him, or, ultimately, the doctor. Yet, it is sometimes hard to decide when the patient is really unable to participate in decision-making, and, furthermore, when (and whether) the preferences he expresses can be entirely ignored. So the question of how to assess the decision-making competence of the patient arises. From the point of view of medical and legal practice, yet another question is important: who is entitled to decide on behalf of a patient who has been assessed as incompetent, and which solutions – institutional, social, or legal – should be adopted to protect the rights of the incompetent? In works on bioethics, this cluster of problems is often termed "the ethics of surrogate decision-making".[1]

Most authors have come up with two general principles, which can be summarised as follows:

– the patient is considered capable of decision-making, unless and until it is completely obvious that he cannot manage this task;

A.-T. Tymieniecka and Z. Zalewski (eds.), Analecta Husserliana LXIV, 263–269.

- even if the patient is incapable of decision-making, the person deciding on his behalf should always take into account – if it is at all possible – the preferences he expressed before losing competence.

Underlying the above two principles is the meta-principle that the patient's own will has absolute priority in all situations in which s/he fulfills the competence criteria. A person undergoing medical treatment, like any free person in any situation, is entitled to decide about everything that concerns him/her. But what should be done if a patient is most probably incompetent to decide, but this is not evident by adopted standards? The paternalist option is one of the possible answers. However, the fear of paternalism, often expressed in bio-ethical writings, is so strong that even in cases where we lack a clear recognition of the patient's incompetence, it is suggested that an obviously detrimental alternative chosen by the patient is better than the imposition of a doctor's decision. Such is the spirit of the law too. One of the reasons for this option is expressed in the slippery slope argument stating that imposing decisions in extreme situations can easily degenerate into a habit of ignoring the patient's will and preferences in situations where there are no doubts about the patient's capability to decide.

In Poland, the situation can sometimes be as follows: on the one hand, doctors, brought up in the paternalistic tradition, tend to disregard the patient's opinions whenever possible; on the other hand, afraid of possible legal consequences, they frequently yield to the patient's will without much reflection.

The above approach makes it impossible to find any satisfactory method of dealing with these most difficult, yet most frequently encountered, cases, where it is quite likely that the patient lacks to some extent the competence to decide in matters connected with his/her treatment. Decisions in such cases are taken rather at random: depending on secondary circumstances, the doctor may take a straightforward paternalistic approach or may leave the patient to his own devices, totally unprepared for making such a decision, and possibly confused by the contradictory statements of various specialists.

Any decision-making process can be presented as follows:

$$G \underset{K}{\rightarrow} A \underset{VK}{\rightarrow} < A, \{U_i\} > \underset{?}{\rightarrow} < A, a_0 >$$

Where G is the goal whose realisation is desired by the subject, the set A consists of alternatives of which the subject is aware, U_i is the evaluation of A's elements, K is the knowledge that makes it possible to distinguish between those potential acts that are conducive to G and those that are not, V, the accepted system of values, and a_o, the decision taken.[2]

A person is considered competent to decide if s/he is able to reason in accordance with the above scheme. Competence does not need to be understood as an absolute ability, it should be judged rather in relation to particular decision-making situations. This means that one and the same person can be competent to make some decisions, but not others.[3] It also seems plausible that competence has degrees: that a person can be more or less competent in certain matters. Linguistic usage seems to confirm this view: person A can say that person B is more competent in certain matters than s/he is, yet this does not mean that person A is completely incapable of making any sensible decision in these matters.

If we can speak of degrees of competence, we can also distinguish between two kinds of incompetence: incompetence in the strict sense (total incompetence); and incompetence in a broad sense (partial incompetence). A person's total incompetence (in a certain situation) implies that the whole burden of the decision must be borne by someone else. We can speak of partial incompetence when the person in question satisfies, in principle, the formal requirements of competence, but resorts to other people's help in the actual decision-making. The above could suggest, and quite convincingly, that total incompetence consists in failing to satisfy all of the requirements mentioned before (in particular, the ability to reason, draw conclusions, and possess a consistent system of values), whereas partial incompetence consists in failing to meet only one requirement: sufficient knowledge. However, adopting such a view would be entirely mistaken.[4]

In each case of "expert advice" in the decision-making process, the role of the expert exceeds to a large extent the mere furnishing of factual statements and rational forecasts. Firstly, the expert frequently participates in explaining the problem in question. In many cases the expert has to help define the goals of the decision-maker, who, as psychological experiments suggest, is often driven by contradictory desires. In many cases, especially when the decision-maker faces entirely new problems, he cannot evaluate the situation according to his old system of values. Thus, he changes or develops it, under the influence of the expert – sometimes without either party even realising it. A person emotionally involved in a choice frequently has problems with estimating the probabilities of future events, which can lead either to over-optimism, or to unfounded pessimism. All the above factors make it plausible to speak, in such cases, of the phenomenon of partial incompetence, and in these situations the expert often decides "on behalf" of the person who is, formally, the decision-maker. I have termed this "'quasi-surrogate' decision-making".

Going to a doctor can be an example of seeking an expert's advice in decision-making. The incompetence of patients is not only observed in extraordinary cases. The ability to make decisions, in the case of any serious illness, is at least to some degree and for some periods, limited. Serious illnesses involve the danger of death, a change in the patient's whole lifestyle, or essential alterations in his/her system of values, which, in turn, force the patient to rethink his/her attitudes and preferences. Not infrequently, serious illnesses also call for a non-standard treatment. For the majority of patients who suddenly realise the threat posed by a serious disease, the shock brought on by the new situation, the lack of sufficient knowledge, prejudices, problems with adequate evaluation, and the fast pace of changes, render them incapable of making rational decisions. They are just not competent enough. Thus, we can say that in such situations the doctor must to some extent take responsibility for decisions made "on behalf" of the patient, or at least for influencing the patient's choices. This by no means involves the need to "manipulate" or embrace the paternalistic attitude. In the vast majority of cases the doctor him/herself is not an expert at decision-making, and resorts to a psychologist's or psychiatrist's help; however, as far as the knowledge factor is concerned, s/he can typically trust his/her own judgement and intuition. Sometimes, the doctor is also emotionally involved, has preferences and prejudices of his/her own, and, not infrequently, his/her own concept of what is good for the patient. This results in a variety of goals, values, elements of knowledge, prejudices, preferences, forecasts, and evaluations by many persons: the patient, the doctors, the patient's family, legal representatives, etc., intertwined in the actual process of decision-making. Typically, the resulting knot is so tight that the constitutive factors cannot be disentangled. The decision, so to speak, emerges in a dialogue: most of all in the doctor-patient dialogue. And this is the reason why both the paternalistic and the subordinate approach – if the latter is based on respecting the patient's freedom and trusting that s/he knows best what is good for him/her – are grossly one-sided.

If there is a grain of truth in the above remarks, three main conclusions follow: Firstly, the doctor should assume that the patient is always partially incompetent. Thus, secondly, the doctor should not rely entirely on the patient's own will. Thirdly, the doctor can follow the above two rules without embracing the paternalist stance, if s/he strives to enhance his/her own competence as well as the patient's (e.g., by gaining better knowledge of the patient's situation, his system of values, motivations, etc.). At this stage, it is worth examining the standards adopted in cases of totally incompetent persons, i.e. cases when, in our terminology, "quasi-surrogate decision-making" becomes "surrogate decision-making".

In surrogate decision-making, four standard methods of reaching a final decision exist: 1. evaluation of the "best interest" of the patient, 2. a "reasonable-person" standard, 3. an "identity of interests" standard, 4. a substituted-judgement standard.[5]

From the theoretical point of view, the above methods touch on many ethical issues: ethical relativism, rationality of attitudes and preferences, the ability to evaluate objectively, and finally, questions of what is good for a human being and what his/her ultimate goals are. Let us, however, leave these aside, and concentrate on the practical side of the four methods. An expert would tend to identify at least the first three: as someone trained to deal with such situations (standard method 2), reasoning by analogy, "this is what I would do, if I were him" (standard 3), the expert thinks he can best define "the best interest" of the patient (standard 1), and thus find the appropriate solution. The fourth method is the most problematic, as it is usually not known what the patient's decision would be were he capable of making one. The patient probably never expressed his/her wishes, and even if s/he did, when we take into account the essential changes in his/her situation, we can never be sure whether s/he was then *competent* to make the decision. Thus, the question arises as to whether the patient's preferences were "appropriate". On the other hand, if we admit that it could have been otherwise, then the fourth method may lose its appeal altogether. The temptation to identify all four also has a theoretical dimension: within a consistent theory of rational attitudes and preferences, all the methods should lead to the same conclusions. However, since in a real situation we never know all the premises necessary for decision-making, we should never *identify* the four, neither should we rely thoroughly on only one. In practice, the four methods will not lead to one-sided choices only if they are applied *independently*, as complementary and probing each other. Only with this assumption can the four be employed in cases of partial incompetence. Again, a reference to the doctor-patient dialogue is valid, for only in this dialogue can we work out an understanding of what is good for the patient, define shared standards of rationality, create mutual empathy and try to grasp the most essential – though often subconscious – reasons for the choices preferred by the other person. This, in turn, enables us to recognise how deep the incompetence of both parties is, and which elements of the situation it concerns, and, consequently, can lead to an enhancement of the competencies of both the doctor and the patient. Only in such a dialogue can a common decision be worked out. The responsibility for the whole process lies with the doctor, and it is in this sense that we can speak of him making the decision "on behalf" of the patient.

The process of decision-making in the doctor-patient dialogue turns out to be a highly complicated matter. Not only does it involve intertwining the decision premises of several persons, not only do these reshape and change dynamically in the process, but the methods of reaching practical solutions are far from unquestionable as well. Firstly, the distinction between total and partial incompetence (and the concept of degrees of competence) can be objected to. No sensible criteria have been put forward that would enable us to define the degree of competence, although an experienced psychologist or psychiatrist can, most probably, make the necessary distinctions in practice. Within decision theory, we can also speak about premises being in closer or looser accordance with reality. Disregarding the theoretical hurdles, this seems, at least to the author, to be a promising way to approach the problem of "paternalism *versus* leaving the patient to his own devices".

Secondly, doubt can be cast on the idea of applying the four methods of dealing with cases of total incompetence together to work out consistent common ground for decision-making, in cases of partial incompetence. However, again, it seems that this can contribute to a more precise and comprehensive description of the decision-making situation. The proposed way of decision-making resembles to some extent the psychoanalytic approach. Here the next problem emerges: *time*. Time is needed to get acquainted with the patient, time is needed to teach him/her the fundamental principles of medicine, time is needed to convince him/her and, most of all, to calm him/her. Time is needed, as well, to get ready to make a decision on the patient's behalf. And very often, there is *no time*. Sometimes this is due to organisational inertia, sometimes to the lack of the *habit* of devoting more of the doctor's precious *time* to one patient, but sometimes as well to the specific features of the situation in which a decision must be made. What compensates for the lack of *time* is the doctor's *experience*. This turns out to be the most important factor in real situations, where a real human being, whom the doctor is taking care of, is concerned. This experience cannot be gained except by learning in close contact with another human being.

Kraków

NOTES

[1] Perhaps the most comprehensive account of these can be found in A. E. Buchanan and D. W. Brock, *Deciding for Others. The Ethics of Surrogate Decision Making*, Cambridge University Press, 1989.

[2] K. Szaniawski, "Philosophy of Decision Making", *Acta Psychologica* **45** (1980), pp. 327–341.

[3] This is the option of A. E. Buchanan and D. W. Brock, see ibid. pp. 59–65.

[4] See K. Szaniawski, "Sociology and Model of Rational Behaviour", *Dialectics and Humanism* **2** (1974), pp. 133–144.

[5] See e.g. Ruth Macklin, *Mortal Choices*, New York: Pantheon Books, 1987, pp. 99–111.

ANNA ALICHNIEWICZ

JUSTICE AND BENEFICENCE IN DEFICIENCY MEDICINE

As a matter of fact, the title ought to end with a question mark rather than a full stop, because the subtitle of my paper can be formulated as a hermeneutic approach towards the patient's needs and wants.

Justice – one of the four main principles of medical ethics or bioethics – has become even more important and more often discussed recently. It is, first of all, the distributive meaning of justice (using the classical Aristotelian typology and distinguishing among three orders of justice)[1] which makes the interest in justice so permanent in bioethics. Moreover, we can agree with J. Fletcher, who says: "It seems that the heart of the problem of distributive justice is how we are to allocate our resources".[2] Why has this issue become so urgent?

The main reason for that is perhaps a rather strange phenomenon presently observed, that in fact every health care system meets the main criterion of deficiency medicine. Not only economically backward countries but also the very rich ones are in the situation that, although often enormous and ever-rising sums of money are devoted to cover the costs of health care,[3] many patients cannot obtain the medical help they need. With large expenditures directed to the medical service without being able to finance it sufficiently, a supposition arises that the resources are allocated improperly. If this is the case, the opinion occurs that the deficiency has resulted mainly from the wrong health policy which is caused by the unjust health care system. In this way we have returned to the question of justice.

Of course, the notion of justice is too vast to serve as a theoretical basis for the health care system's construction on its own terms. Even if the traditional definition of justice as "giving to each his due"[4] could be interpreted as treating like cases alike,[5] what we have given is still just a minimal or formal principle of justice. There is no specification of what kinds of respects ought to be taken into account. This lack makes possible many interpretations and it is, all the same, a reason for developing different and rival theories of justice. Actually, if we look into bioethical literature on the problem of health care delivery, we can encounter a wide range of theories of justice such as libertarianism, utilitarianism, egalitarianism, the Rawlsian Maximin Contract Theory, to mention the main ones. The theories have been

A.-T. Tymieniecka and Z. Zalewski (eds.), Analecta Husserliana LXIV, 271–281.

diligently and precisely analysed and discussed by bioethicists who deal with the problem of medical care resource allocation. These ethicists have also taken part in social debates upon existing national health care systems.[6]

In "Cross-Cultural Perspectives in Medical Ethics" we can read: "Each of the theories of justice under consideration offers a theoretical basis for answering some basic question concerning justice in health care. We have seen, however, that none of them provides unambiguous answers to all of the questions and that each depends for its application upon a wealth of empirical premises, many of which may not now be available. Each theory does at least rule out some answers and each supplies us with a perspective from which to pursue issues which we cannot ignore. Nonetheless, almost all of the work in developing an account of justice in health care remains to be done".[7] I fully agree with the statement that there remains so much to be done with the problem of justice in health care, but I am not convinced that the perspective offered to create a just health care system is really a sufficient way to achieve this goal. What is proposed as an adequate method for dealing with this task is an application of general theories of justice to solve the problem of justice in medical care. Medical ethics or bioethics dealing with the whole problem of just medicine is taken for granted as applied ethics.

Such a conception of medical ethics, although still dominant, has been criticized recently. Recent years have contributed a critical account to the model of medical ethics which emphasized the rights-duties framework in the making of medical decisions, and have attempted to rethink the virtue-value and rules-moralities relationship in medical ethics.[8]

The primacy of the principle of autonomy which determined the climate of bioethics in the late sixties and seventies was not only an intellectual result of the Enlightenment, but also a reaction to a long history of medical paternalism, particularly since the technical and scientific power of modern medicine made the paternalistic model impossible to accept. The absolute, unconditional respect for the patient's autonomy was recognized as a much more adequate answer to the situation of modern medical treatment, with its abilities and possibilities that had not been even envisaged by the earlier ages of medicine. Nevertheless, the further development of technology-oriented medicine led to the new dilemma. The scientistic ideology not only produced medicine regarded as a kind of technology, but it also created a specific attitude towards its genuine task, and such a meaning of medical goals was adopted by medical professionals and by patients alike. The scientistic ideology with the Promethean Myth in its background promised an unlimited progress of mankind. Evil was regarded as an accidental and temporary stage in world history. It also concerned the kind of evil which is so commonly recognized and experienced by humans – disease and death. Thus medicine considered technologically was believed not only to be able to cure diseases successfully but, what is most important, even to fight and conquer death. The modern dream of a world free from disease and death has been transmutated into the medical everyday-life phantasmagoria.

The above belief has determined the patient's hopes and wishes and it is probably responsible for the increasing spiral of demands which are directed towards doctors. It is also responsible for the common view of medicine as an area where miracles often occur. Many people (some doctors included) treat medicine as a scientific, of course, but also "a miraculous" discipline of knowledge. This view can be dangerous in its results and it must be differentiated from the very reasonable treatment of medicine as also a kind of art, which refers especially to clinical medicine and the healing relationship (which I will return to later).

Eventually, patients have become confused – they often cannot distinguish between their virtual needs and unreasonable wants. The latter have often taken the place of the former. On the other hand, the technological option has misled doctors as well. They have also begun to view their role unrealistically and regarded the fight with diseases and death at any price as a genuine goal of medicine. Their treatment possibilities made possible by medical high-technology caused the necessity of an autonomy-based medical ethics with its emphasis put upon the patient's right to express his/her wishes and to self-determine his/her good or his/her best interest. The doctor's duty has become to respect the autonomy of the patient and to fulfill the patient's wishes. It has had obvious positives and advantages for the doctor-patient relationship, which has become that of partnership and respect, at least on the physician's side. Nevertheless, it has also had its weaknesses. The proper idea of informed consent has often drifted into the situation when the whole weight of decision-making has been shifted onto the patient, whose wishes supported by technological possibilities might have been unreasonable. Combined with modern physicians' technical abilities and skills, accompanied by their awareness of their duty to respect the patient's will, these wishes have begun to determine medical treatment which could have been futile or even medically unreasonable. The described phenomena have largely contributed to modern medicine becoming one of deficiency, notwithstanding how well-off the country is. Paradoxically, the better economic situation could make the situation even worse. The richer the country is, the more rapidly the spiral of demands and medical costs increases.

The above could be recognized as an important cause of the next transformation of the bioethics attitude. The principle of autonomy has been replaced by the principle of justice. The good of the patient, which – although defined differently – prevailed in both medical paternalism and medical respect for patient's autonomy, has had to be balanced against the common good. The latter has happened to be defined as the good of other patients, medical personnel, medical institutions or society as a whole.

The last transformation in medical ethics priorities has carried a great danger for itself. It seems to be very erosive and it can lead to the complete annihilation of the very nature of the medical relationship as a healing relationship based upon the patient-doctor bond. An answer to this impediment can be considered an attempt to restore beneficence as the main and guiding principle for medical practice.[9] This option has also met with great interest among bioethicists dealing with moral issues concerning health care systems and health care costs. In their book *Medicine and Money,* F. H. Marsh and M. Yarborough expressed their conviction that beneficence can serve as the best solution to the problem of justice in health care.[10] According to the authors, the implementation of beneficence as a basis for health care will solve the problem of justice. It will guarantee health care to be just. Thus just health care is the doctor's practice led by the principle of beneficence. These bioethicists have not shared the opinion, which is quite common nowadays, that none of the principles can or should be taken as absolutely dominant and always overriding.

Yet critics have gone further. Among many weaknesses that have been pointed out about the applied medical ethics model, the most important one is not the problem of the balance of principles as such. The crucial criticism has questioned the whole conception of medical ethics as an applied ethics. The applied medical ethics model seems to neglect the fact that medical care is an ethical practice itself, and, as such, does not need such an external ethical framework which is said to have operated and to operate as a *forma operans extrinsecum* for medical problems *materia*.

Thus, the point is not the fact that in the post-modern pluralistic society there is no agreement about moral theories and that no doctrine or system governs universal consensus. Neither it is the point that we do not know which moral principles or rules to apply, although this uncertainty can lead to a rather dangerous conviction that as there are various plausible yet conflicting theories, each medical decision will be condemned by some of them and accepted by others.[11] Yet the most important is the assumption that "medicine is intrinsically a moral activity since it has a unique character as a healing relationship between doctor and patient", as we can read in *Changing Conceptions of Medical Ethics* by Henk Ten Have and Gerrit Kimsma.[12] For this approach, a basic view is a conviction that medical ethics should be conceptualized rather as reflective practice than as applied theory.[13]

My view is to replace the application of the principles through intrinsic revealing of patient's needs and wants. Why should I approach it that way?

Let me begin my explanation by referring back to Marsh-Yarborough's conception. What is it that the restoration of beneficence in medical ethics made so attractive for them and so promising regarding the issue of justice? We ought to look at how E. D. Pellegrino and D. C. Thomasma defined beneficence in their famous book, *For the Patient's Good*. They said: "In fact, we regard beneficence as a shorthand for 'best interest', which includes both respect for autonomy and a benefits-to-burdens calculus about the quality of a person's life. Thus, best interest conflates several important realities and is based on neither paternalism nor autonomy".[14] And then: "The doctor's first duty is to give the patient sufficient information, not primarily for the sake of an 'informed decision', which is very difficult in the face of various risks and uncertainties, but to help the patient to discover and express his or her own values".[15] Yet, first of all, it seems that, in fact, it is the so-called covenantal model of the doctor-patient relationship which makes the theory so attractive for bioethicists dealing with health policy problems. According to Pellegrino and Thomasma, the covenantal model is opposed to the Lockean contractarian model which is formed by two or more autonomous entities as a bond for some mutual good. In the covenantal model the bond is based upon the need of only one party.[16] It is need which seems to be the crucial idea. And, in fact, this is the point Marsh and Yarborough have referred to. They have underlined that the patient comes with needs which require the doctor's answer. Thus the patient-doctor relationship is "an ontological bond of need and response".[17] Thus the real problem is how the patient's needs can be discovered.

As I have mentioned, in the new approach the basic assumption is that medicine is intrinsically a moral activity, inherently a moral profession, using Eric Cassell's expression,[18] while ethics can be considered as the hermeneutics of moral experience.[19] The moral experience of the patient is what initiates the patient-doctor bond. The patient presents his/her symptoms, complains of his/her pains, disabilities and discomforts. In that way, the patient begins his/her story because each physical symptom is greatly influenced by his/her biography. It follows from the trivial statement that none of the symptoms can be traced backwards to some physical deficiency.

Man is also the lived body, and the sharp Cartesian division into the psychic sphere and the somatic sphere is an artificial simplification. Each physical experience is at the same time the psychic one and vice versa. Therefore, each patient's complaint is impregnated by the whole of his/her being. Besides, following Heidegger, man is *In-der-Welt-sein*. The traditional or even applied ethics approach has tended to isolate man from his situation in the world, to render the patient as an isolated individual, ahistorical, asocial, acultural. This is a most falsifying option of the human existential condition, and besides it also darkens the image of the patient's illness and his/her needs. The illness, or strictly speaking the way of experiencing it, and its impact for the patient and therefore the needs directed towards the doctor are co-defined by the world context, of which medicine is also a part.

This aspect covers but half of the problem. The remaining part is the doctor's potential of medical sciences predetermined by his cultural background with its hierarchy of values, abiding beliefs and convictions. It is doubtless that the disease appears differently in the consciousness of the patient and that of the doctor and thus it functions differently for them. Yet it is the doctor's revealing of the patient's apprehension of his/her illness that is the necessary condition for the real healing relationship. From the point of view of hermeneutics, the patient is regarded as a kind of text the doctor has to reveal, understand and interpret. By the very nature of his profession, the doctor seems to be summoned to the hermeneutic task, which, as stated by Gadamer, consists in the understanding approach towards the other and in the interpreter's eager and attentive listening. The interpretation of the patient-as-text is rendered possible owing to the universe of understanding they both take part in, that is, the so-called hermeneutic circle.

This approach seems to be especially suitable for medicine also because it "abounds in stories";[20] it is an entirely narrative domain. Patients have told their stories to doctors since medicine began to exist, that is practically forever in our culture, as medicine is one of the oldest sciences and it arose from the former ancient art of healing. Doctors would tell stories about their patients as well, and these stories used to be an important part of their teaching of medicine, communication with colleagues, conveying information and developing the discipline.

This narrative character has been lost (at least declaratively) by modern medicine, which, trying to meet methodological criteria, discarded stories and replaced them with ideal abstract types. Thus, in modern medicine storytelling became an "anecdotal medicine" and as such is often regarded as "antithetical to modern scientific medicine".[21]

Such a view seems to be based on the misunderstanding and the confusion of the two notions of medicine: medicine as a science and medicine as healing (medical cure and care). The first one has to meet criteria of empirical scientific research, so it deals with generalizations and ideal models of pathological processes and diseases. Scientific knowledge of these ideal models and general types is, of course, vital and necessary for a proper healing activity, but it is not sufficient. The doctor always treats a particular patient "in particular circumstances at a particular moment in time".[22]

Moreover, as a science, medicine has adopted the view that it should be value-free, which is absolutely improper and even contradictory when the healing relationship is concerned. As a matter of fact the whole area of medical care is impregnated by values: values of a given patient, values of a given doctor and values of the society and culture. Terms like health, normal state, pathological state, patient's best interest, goals of curing and caring, that are basic in every healing relationship, are deeply value-laden. Values are inherent to medicine as healing just as they are inherent to narratives.

Thus the character of healing as a narrative domain should be restored. The story the patient unfolds may be fragmentary, as narrative always is, may be in many points vague and incomplete. The patient as a text has his/her main meaningful matrix to be revealed, but the doctor always deals with uncertainties and unknowns, because there is always "the struggle between diversity and structure".[23]

Since the patient can be regarded as a kind of text to be interpreted, theories and methods of interpretation of other texts, especially literature, are very valuable for this task. It is certainly impossible to demand that doctors be skilled enough in such theories and practice, but we can suggest their just reading novels. This can be very helpful, as literature gives us a number of archetypal stories, and, using Ellos' expression, "there is a kind of deeper knowledge and experience of some primal stories being told over and over again".[24] Moreover, Kay Toombs shows what important cognitive input can be given to the doctor by fiction, where the meaning of the illness, its symptoms and the state of being ill are rendered to the patient. It is in literature that doctors encounter diseases as they appear to patients and as they are experienced by patients. Reading novels, thc doctor can learn what the phenomena known to him/her from the medical point of view mean to the patient.[25]

The narratives developed by doctors who have been ill and especially who have stayed in hospital and then have provided descriptions of their illness experience have confirmed the accuracy of such an approach. It can help doctors to develop their hermeneutic skill and to understand the messages conveyed to them by patients.

Every activity of the patient becomes his/her self-presentation and has a specific meaning. It seems to be obvious as long as a verbal activity is concerned: words, used expressions, and specific language play an important part in creating the meaning of the narrated text. Yet some feelings and emotions cannot be verbalized, or the patient cannot or does not want to do it. Nevertheless, the essential and significant part of the patient's story is given through emotions and feelings. Thus the doctor should be very sensitive to each non-verbal message like gestures, facial and body expressions and the emotional climate as a whole.

Nothing can be interpreted apart from the whole hermeneutic situation, as everything is like a particular place within a given structure, i.e. it is a patient in his/her situation who determines a specific, unique meaning to be revealed. It is indispensable to reveal the personal myth of the patient. "Each of us carries and lives out a personal myth", says Ellos,[26] and this myth, even if it is shared with many others, gives significant dimensions to the structure of one's personal experience.

As the hermeneutic task of the doctor includes attentive listening and understanding, and asking questions as well, his/her ability in listening and asking proper questions should be developed. Doctors should be skilled both in story listening and storytelling, as the task of interpretation involves both activity and passivity. Doctors should also be aware of the fact that asking questions is a difficult art, since there is always a danger that you may pre-determine answers and in this way reshape and deform the pattern you are to reveal.

Every text, or at least every good one, is so complex and multilayered that it is practically unlimited, and so its interpretation is a never-ending journey and its reinterpretation is always possible. While searching for the meaningful and meaning-giving pattern, the doctor should realize that his/her task is first of all to serve towards finding of a solution for the present and the future of his/her patient.

However, the patient-as-text contains the mystery of man's being and as such is inexhaustible. It is a never-ending story, never to unfold itself fully, no matter what the patient's narrative abilities might be. R. M. Zaner says: "Every life is linguistically inexhaustible, there is always a richer tale to be told that can never be wholly captured in words, no matter how evocative they might be".[27] The non-existence of a single definitive and comprehensive interpretation is acutely marked in this case. Thus the doctor is condemned to uncertainty. Yet the text displays many levels and, even though many of them remain beyond the scope of narration and interpretation, we can assume that as a whole it is organised, it constitutes a given pattern, ambiguous as it might be. Nevertheless, we can also assume that the pattern renders some leading meanings marking the patient's biography, his/her aims and therefore needs. The needs appearing in the doctor-patient relationship may vary. In the majority of cases it means cure, though that is not always possible. Yet restoring soundness to the body is but a part of healing. Healing may be considered as the restoration of the patient's integrity, the patient's wholeness, either his/her former state or just the achievement of a better one. Healing may be also considered as the restoration of autonomy, or making possible a particular lifestyle. The notions known from applied medical ethics appear at this point, yet let us remember they are approached quite differently from those in the application model: "From a hermeneutic perspective, virtues and principles are not independent of interpreters; they are not the precondition of interpretation but its result".[28] The doctor has to compromise medical indications with the patient's preferences in a way which enables the patient's needs to be revealed. What is very important here is the fact that the doctor does not adapt any external system of norms and values to the situation of the patient but rather tries to reveal moral dimensions intrinsic to the situation and values and norms which are inherent and which generate the meaning of the illness as experienced by the patient. The perspective for doing this must be that of the insider, because all the phenomena occur within the doctor-patient relationship, so their extraction from this context will change their sense. Thus it is assumed that the doctor will not be able to grasp the patient's genuine needs if he takes the external view.

Healing is here understood as mutual, dialogic hermeneutic effort, the doctor and the patient revealing the structure of meanings marking the sphere of the patient's sense of life and therefore determining the patient's needs. Thus the aim of interpretation is to discover the patient's needs as something following from the meaningful pattern of his/her life revealed and understood in this process, and to distinguish them from the wants which have not been contained in the revealed sense of the patient-as-text. The needs constitute an integral part of this structure, whereas the wants become as if an accidental addition, or something that is excessive or even downright contrary to the meaningful pattern.

Now, I would like to make some additional remarks.

Firstly: it is the patient who is able to come into the intellectual and emotional contact with the doctor I have talked about. I have not distinguished between a competent and an incompetent patient. I have not even used this notion. Of course the greater the patient's ability to tell the story of himself/herself, the more likely that the hermeneutic task will be fulfilled. Yet, as a matter of fact, also in informed-consent ethics it is a threshold competence that is referred to. The threshold competence requires the patient to be treated as autonomous and able to make decisions about himself/herself. And full informed consent is only a theoretical ideal.[29]

In the hermeneutic approach also, of course, the more the patient exceeds the threshold competence, the more fruitful the hermeneutic bond with his/her doctor can become. Yet even in case of an incompetent patient (for example: the patient in coma, in PVS, debilitated or retarded), the proxies should be involved in the hermeneutic task on behalf of the patient. There is an important difference which occurs here. In the case of the patient who used to be competent in the past, the doctor has to focus primarily on the patient's former attitudes and try to reveal the patient's own meaning of his/her life, treating storytelling unfolded by proxies as intermediate. In case of, for example, severely handicapped newborns, it is their surroundings, first of all the parents' attitude and life context, that mainly have to be taken into account.

Secondly, my attempt has been to find a new view of the problem which is conceptualized in the applied medical ethics notions as a problem of justice in medicine regarded as a clinical practice. The doctor who cures and cares for an individual patient cannot be expected to take general issues of national health policy like resource allocation or health care financing into account. That is the problem which health economists deal with. Policy-makers constitute an institutional framework within which doctors have to perform their task of healing in an ethically proper way. For example, the resource allocation problem I began with may be solved on two levels: the general or macro one where health economists and health policy-makers decide on priorities in resource allocation, and the individual or micro one where the doctor decides about using given resources according to his/her patient's needs.

Medical University of Lodz

NOTES

1 Aristotle, *Nicomachean Ethics*, Book 5, Chapters 1, 2 & 3.

2 Joseph Fletcher, "Ethics and Health Care Delivery: Computers and Distributive Justice," in: Robert M. Veatch, Roy Branson (eds), *Ethics and Health Policy*, Ballinger Publishing Company, Cambridge, Massachusetts, 1976, p. 102.

[3] Frank H. Marsh, Mark Yarborough, *Medicine and Money*, Greenwood Press, New York, 1990, p. 9.
[4] Albert F. Jonsen, André E. Hellengers, "Conceptual Foundations for an Ethics of Medical Care," in: Robert M. Veatch, Roy Branson (eds), *Ethics and Health Policy*, Ballinger Publishing Company, Cambridge, Massachusetts, 1976, p. 29.
[5] Tom L. Beauchamp, Le Roy Walters (eds), *Contemporary Issues in Bioethics*, Wadsworth Publishing Company, Belmont, 1982, p. 33.
[6] Nancy S. Jecker, Eric M. Meslin, "United States and Canadian Approaches to Justice in Health Care: A Comparative Analysis of Health Care Systems and Values," *Theoretical Medicine*, 1994, 15, pp. 181–200.
[7] Allen Buchanan, "Justice: A Philosophical Review," in: Robert M. Veatch (ed.), *Cross-Cultural Perspectives in Medical Ethics: Readings*, Jones and Bartlett Publishers, Boston, 1989, pp. 294–295.
[8] *Vide* e.g.: Alisdair MacIntyre, *After Virtue*, University of Notre Dame Press, Notre Dame, 1983; *Whose Justice? Which Rationality*?, Gerald Duckworth & Co., London, 1988.
Edmund D. Pellegrino, David C. Thomasma, *For the Patient's Good. The Restoration of Beneficence in Health Care*, Oxford University Press, New York, Oxford, 1988.
[9] *Vide* E. D. Pellegrino, D. C. Thomasma, *op. cit.*, *passim.*
[10] *Vide* F. H. Marsh, M. Yarborough, *op cit.*, *passim.*
[11] Henk ten Have, "The Hyperreality of Clinical Ethics: A Unitary Theory and Hermeneutics," *Theoretical Medicine*, 1994, 15, p. 117.
[12] Henk ten Have, Gerrit Kimsma, "Changing Conceptions of Medical Ethics," in: U. J. Jensen, G. Mooney (eds), *Changing Values in Medical and Health Care Decision Making*, John Wiley & Sons Ltd. 1990, p. 40.
[13] Henk ten Have, *op. cit.*, p. 115.
[14] E. D. Pellegrino, D. C. Thomasma, *op. cit.*, p. 20.
[15] E. D. Pellegrino, D. C. Thomasma, *op. cit.*, p. 140.
[16] E. D. Pellegrino, D. C. Thomasma, *op. cit.*, p. 6.
[17] F. H. Marsh, M. Yarborough, *op. cit.*, p. 72.
[18] Eric J. Cassel, *The Healer's Art. A New Approach to the Doctor-Patient Relationship*, Lippincott, Philadelphia, 1976, p. 109.
[19] Henk ten Have, Gerrit Kimsma, *op. cit.*, p. 46.
[20] Eric J. Cassel, *The Nature of Suffering and the Goals of Medicine*, Oxford University Press, New York, Oxford, 1991, p. 203.
[21] *Cf.* 20.
[22] Eric J. Cassell, *op. cit.*, p. 179.
[23] William J. Ellos, *Narrative Ethics*, Avebury, Aldershot, Brookfield, 1994, p. 6.
[24] William J. Ellos, *op. cit.*, p. 114.
[25] S. Kay Toombs, *The Meaning of Illness. A Phenomenological Account of the Different Perspectives of Physician and Patient*, Kluwer Academic Publishers, Dordrecht/Boston/London, 1992, p. 107.
[26] William J. Ellos, *op. cit.*, p. 14.
[27] R. M. Zaner, *Ethics and the Clinical Encounter*, Prentice Hall, Englewood Cliffs, 1988, cited by: Henk ten Have, Gerrit Kimsma, *op. cit.*, p. 47.
[28] Henk ten Have, *op. cit.*, p. 124.
[29] E. D. Pellegrino, D. C. Thomasma, *op. cit.*, p. 10.

IGNACY S. FIUT

THE MEDIA AND HEALTH

The objective of my considerations is to investigate the ways in which the media influence human, i.e. public, health. To accomplish this task, I will present some popular theories of media impact and functioning, juxtaposed with the preferences of "the value of health" shared by the media public. Then, some philosophical remarks on the directions of media impact on "subjective health" and "objective health" will be presented, including the differentiation between individuals and social entities, following the assumption that health, being good, is one of the most important and universal human values.

THE THEORIES OF MEDIA IMPACT

The theories of media impact have been prone to evolution for the last fifty years in accordance with the subsequent findings on the real media influence exerted on their addressees' preferences and behaviour. One can distinguish between three theories of media effects: the theory of unrestricted effects of the media, the theory of restricted effects of the media, and the theory of limited influence.

The theories of media effects above all try to explain the basic media functions, which are:

1. Permanent observation of the environment in order to have constant information, making the public aware of changes in the world as well of as their consequences.
2. Providing entertainment and organising leisure time, by offering a profusion of intellectually and emotionally amusing games.
3. Helping people to know and interpret everyday events, and providing them with synthetic layers in the shape of a certain, possibly homogenous, entity.
4. Providing individuals with information useful for active involvement in social life, thus participating in their socialisation.
5. Performing possibly systematic public education on intellectual and social levels.
6. Observing state legislative and administrative bodies, thus fulfilling the function which stems from the thesis that the media constitute a "fourth power" in a democratic society.

A.-T. Tymieniecka and Z. Zalewski (eds.), Analecta Husserliana LXIV, 283–292.

Consequently, the media, according to all six functions mentioned, can strengthen, make dysfunctional, or damage both individuals and a whole society. It can also happen that certain media events may have an equally functional and dysfunctional nature. Presenting acts of terror or deception can cause sympathy, satisfaction, and can be a remedy, but can also arouse fear, admiration, apathy, frustration, aggression, etc. Media influences can best be seen in special campaigns, for example, organised during elections, to fight criminals or corruption, to support homeless people and animals, to promote "natural childbirth", to decrease smoking, or to organise leisure time in an active way, etc.

All of those "combined media activities" aimed at propagating certain ideas are directly or indirectly based on certain standards which are consequently promoted as deserving approval (or rejection) by the public.

In creating media theories, the communications theoreticians aspire to investigate the ways in which the media influence people, by looking for answers to the following questions: "Who is speaking? What is said? To whom is it said? And with what effect?" Thus, the aim of those theoretical approaches is to know about creating a proper message which should bring about concrete, direct, desired and foreseen responses.

One of the first theories of media impact, created as a result of positive media experiences during the First and Second World Wars and attributed to the development of advertising, was the "Bullet Theory". This theory proclaimed that the media exerted great power over society (L. M. Walters, 1995). It said that people are vulnerable to media messages which, like bullets, can hit the addressee and make the public do exactly what the media planned (Schramm, 1971). The rationale for the "Bullet Theory" was behavioural psychology, where treating a message as a stimulus causes a very concrete response. This theory assumed, however, that the public was very naive and homogenous. According to evolution and the development of the media and society, this theory is only slightly applicable now. Instead, the theory of restricted effects exerted on the public by communication has been introduced.

The theory of restricted media effects claimed that the media were only one of the factors shaping public opinion. Media messages seemed to be themselves shaped by other factors such as perception, brightness, understanding, processes typical for different social groups and group norms, opinion leadership, etc. (Klapper, 1960). Interpersonal communication also plays an important role in the transformation of media messages. It strongly influences people's decision-making. Therefore, researchers pointed out that a media message should first reach the opinion leaders who then reshape it and pass it on to other people, exerting influence on them, opinion leaders being, for instance, prominent members of political parties, trade unions or the clergy. This theoretical approach was called "a two-step flow of information". In subsequent research it was discovered that the flow of information can actually have more than two steps, that the opinion leaders frequently change, and that there is information feedback between the leader and his or her addressee (Lin, 1971). These observations made theoreticians formulate the concept of a multi-level information flow, which developed into another version, the "diffusion of innovations theory".

This theory concentrated on the ways in which people make up their minds when looking for new ideas, actions, and products, and how those ideas develop in a social system. According to this approach, media messages and interpersonal information flows are equally important for spreading innovations in a society. The media import those innovations, and interpersonal communication channels make individuals accept them. Thus, the combination of the two, media and interpersonal channels, creates the most effective way of changing the behaviour and preferences of the public (Rogers, Shoemaker, 1971).

The next concept of media influence on the public mentioned was the theory of limited influence. This one claimed that obtaining information goes hand in hand with its usefulness, and with such matters as satisfaction, holding to cultural norms, and agenda-setting. This concept drew the attention of researchers, away from the sender and towards the information receiver. It was observed that a message's usefulness, the receiver's interest, the entertainment value, and individual characteristics differentiate the openness of an audience toward media messages. A good example of the above is the strategy of Jerzy Owsiak's action "The Orchestra of Festival Help" (Orkiestra świątecznej Pomocy), where those factors played a major role. The grounds for this kind of theory of influence are the evolutionary theory of knowledge, which assumes the gradual and strengthening imprinting of a human being's epistemic apparatus onto the goal of action (K. Lorenz; I. S. Fiut, 1994).

Consequently, Elihu Katz pointed out that investigating media effectiveness should concentrate on the question: "What do people do with media?" (Katz, 1957). According to Katz, Blumler, McQuail et al., the public makes an active selection and use of a media message in order to fulfil certain intellectual and emotional needs (J. Grzybczak, 1995). This model is usually called the theory of uses and gratification. However, this theory became subject to severe criticism, mainly because of the fact that it operated most of all at the level of collecting empirical data, and thus did not have any real prognostic value. Indeed, it closely ties media functioning with public tastes and preferences.

The next theory aiming to replace the previous one was Melvin De Fleur's theory of cultural norms. It said that the cultural norms of a given public determine its patterns of behaviour and preferences. A media message's task, correspondingly, is to launch certain actions according to those norms (De Fleur, 1970). Another version of the theory of cultural norms was McCombs and Shaw's concept called "agenda setting", which underlies a strong and positive relationship between the media, and the importance of all media topics for individuals in the public. Those topics usually correspond to the norms which organise people's everyday life.

However, the media do not entirely mirror the life of the people, but rather filter information and try to shape reality in a selective way, always for the public. Because of this, theoreticians of communication draw attention to other institutions effecting media messages, such as public officers, spokes(wo)men, and public relations specialists. These actors prove to be the primary source of the bulk of media news, simultaneously creating the primary sense and the ideological grounds of this news. This led Gondy to construct the "theory of the process of subsidising information". The process is an attempt to put pressure on the actions and behaviour of other people by controlling their access to, and opportunities to use, the information. This kind of information is subsidised because its source allows the user to buy it at a lower price (Gondy, 1982, s. 61).

In fact, this is a symbiotic model for decreasing the costs of the flow of information, with its symbiosis between journalists and the sources of information. Senders communicate the news which is important to them more cheaply, journalists reach it cheaply and easily and then pass it on to the public. Thus, the information becomes more professional and credible.

However, all of these theories together are still not sufficient to explain the media's influence on the public. According to Noelle-Neumann's research (1973), the more uniform and homogenous the public (i.e., sharing similar thoughts, emotions and cultural values, and belonging to the same social class), the more effective a media message will be.

The theories mentioned are criticised for "mediotropism of the public", a situation where only the public adapts to the media. A complete theory of media influence should additionally have the aspect of "publicotropism of the media": it ought to take into account the situation where the media adapt to the evolution of the public's behaviour and preferences. It also ought to treat the process of mass communication as a game in which two strategies creatively participate: that of sender, and that of receiver.

Nevertheless, the theories of media influence can sufficiently explain different aspects of a possible impact which the media exert on human health. They lead to a conclusion that this impact has a persuasive and manipulative nature. They can have both a good and a detrimental effect on a human's situation.

THE VALUE OF HEALTH IN THE PUBLIC'S PREFERENCES

The importance of the "value of health" for people will be checked according to the research on "key words", or "flag words", carried out in the Press Research Centre of the Jagiellonian University in 1984 and 1989 (H. Siwek, 1985; W. Pisarek, 1992). The findings show that the Polish public has a stable, established approach to this value.

The method of analysing key words was originally invented by Harold D. Lasswell (1965); it concerned the "verbal political symbols", words making the "key symbols", producing the main notions of many political and social myths. According to W. Dieckmann (1969), this concept of a "political myth" consists of three elements: a doctrine, "miranda", and execution rules. By "miranda" he meant those myths which arouse admiration and are based on beliefs. They serve to strengthen the loyalty of individuals in social structures in their group behaviour, and they support an awareness of the group's participation. The negative equivalents of those elements are: anti-miranda, meaning something unwanted, treated with contempt, representing a rival utopia. In a nutshell, they represent negative values.

Walery Pisarek invented the terms "flag word" and "flag notion" corresponding to this theory, which he used for analysing the preferences of the Polish media public. The flag words, according to him, are "words and phrases which denote the most important content in a given text, and are emotionally biased and, as can be inferred, are being used in a persuasive function", i.e., with reference to a given type of value accepted by the public (W. Pisarek, 1992, p. 18).

The questionnaire survey carried out on a sample of 3230 respondents, who were asked to note their attitudes toward the list of chosen terms – media key words typical for given political, religious, and world-view orientations, helped Pisarek to prepare the list of "words of most accepted meaning" and, on the other hand, "words of least accepted meaning".

The word "health" takes the second position on the list of most wanted denotations, having the support of 73.6% of the respondents, which was exactly the same share of acceptance as for "honesty". These two words took first place among Poles. "Health", as the anti-miranda, was noticed by only 0.4% of respondents. "Health" surpassed such miranda as "love" (65.9%), "justice" (65.5%), "agreement" (60%), "freedom" (59.3%), "tranquillity" (48.7%), "honour" (46.4%), "fatherland" (46.4%), "safety" (42.3%), "faith" (42.1%), "care" (39.4%), "responsibility" (36.3%), "patriotism" (33%), "equality" (32.3%), "moderation" (29.6%), "nation" (28.1%), "sovereignty" (21.8%) and "solidarity" (19.5%).

As I already mentioned, "health" as anti-miranda was noticed by only 0.4% of the sample, whereas "faith", for instance, was noticed by 1%, and "solidarity" by 6.4% (W. Pisarek, 1992, p. 20–21). As far as the rest of the aspects of the demographics, sex-age, place of residence, education, financial status, occupation and religious attitudes, it is clear that "health" is the most important value for "unqualified peasants" (77.7%), people of primary-school education (76.5%), people living in the countryside (75.3%), those of more than 60 years of age (74.9%), students (74.4%), women and children (74.1%) each). On the other hand, "health" is the least important for people who are "rich", "highly educated", and for young persons between 14 and 24 years of age (69.6% each), intellectuals (71.0%), "men" (72.2%), "people believing in God" (73.1%), "farmers with private lands" (73.3%) and "people who do not believe in God" (73.9).

The above findings clearly show that the Polish media public, given its slight differentiation, is very sensitive about the value of health treated as miranda of a media message. This means that the value of health in Polish society, similarly to other societies of European culture, is one of the most important.

Consequently, the intentional approach to this value when using it in media manipulations can successively affect the behaviour of the pro- and anti-health oriented. The value can easily be used in media message manipulations in order to achieve different values and behaviours, which are closer to the sender and producer of a media message. A good example of this process is the many advertising campaigns for individual hygienic garments, diapers, cosmetics, consumption products, etc. Thus, the opinion leaders together with the media can form one of the strongest "coalitions": "for" health, or "against" it (W. Pisarek, 1992, p. 22–26).

In analysing the relations of coexistent miranda and anti-miranda, Pisarek, using the dendrogram method, reconstructed five clusters of values preferred by the Polish media public. They are: the cluster of ambivalent values ("new beginning", "foreign capital", "Europe", "pluralism", "solidarity"), the cluster of citizens' virtues and positive features ("moderation", "safety", "sovereignty", "consistency", "responsibility"), the cluster of patriotically etatist values ("patriotism", "honour", "fatherland", "society", "state", "nation"), the cluster of rustical values ("equality", "health", "agreement", "tranquillity", "faith", "care"), and the cluster of knight-errant, or western-movie-sheriff values ("justice", "fairness", "freedom", and "love")."Health", as can be seen above in those "preference relations", is treated by the Polish media audience as a value close to rustical origin. The author, after comparing these findings with the results of a similar survey carried out seven years ago, noticed that no group of values changed places in the choice structure: people still value the missing things most, or those goods which can be lost. Health by all means belongs to the latter group.

THE NOTION OF HEALTH IN THE ASPECT OF MEDIA INFLUENCE

In the philosophical tradition, the notion of health is usually treated twofold. On the one hand, in an individual aspect, health is taken separately from the environment. On the other hand, the health of an individual is treated in the broader context, mostly that of the social or natural environment (C. F. von Weizsaëcker, 1978; J. Aleksandrowicz, 1988). Additionally, apart from the previous distinction, the notions of "subjective health" and "objective health" are constructed (J. Bejnardowicz, M. Sokołowska, 1975; D. Zalewska, 1989).

After combining both dichotomies, we receive the following understandings of health: objective health of an individual, objective social health, subjective health of an individual, and subjective social health. It can also be inferred that objective health, both individual and social, does not necessarily have to go hand in hand with the subjective sense of an individual and social good condition.

People and societies, being in fact in good shape, quite often have the subjective feeling of a lack of health, thus illness, and *vice versa*. This subjective sense of illness, as a persistent state of consciousness, can lead to the subsequent loss of objective health, both among individuals and societies (Aleksandrowicz, 1988; I. S. Fiut, 1992). On the other hand, by increasing their subjective sense of health, ill individuals and societies can regain a good, objective state. So, these phenomena are usually connected with the possibilities of making the media capable of exerting an influence on the condition of individuals and social groups. But this has to be accompanied by a systematic monitoring of social consciousness, one which would enable a coordination of media strategies and their adoption for the conscious and unconscious needs of people and societies (H. J. Schneider, 1992). "Using the media" should be preceded by special education, which would be based on ethical "principles of intellectualism", "principles of moderation", and the ontologically epistemological principle of "hypothetical realism", making the creators of media news fully aware of the current condition of people's health. The lack of "adequate knowledge" on the part of a sender and of a receiver always leads to exaggeration, and "every exaggeration is a crime". In medicine, this kind of a behaviour is called "an iatrogenic fault" (K. Szewczyk, 1994). Analogically, it can be noticed that a similar phenomenon occurs in the practice of media functioning. This can lead to pathological incidents in treating public health which would cause the loss of health of the people exposed to media influence.

The theory of media influence, taking into account the impact exerted by the media on the subjective sense of health, should mostly follow this kind of media game logic played with the public, where the two strategies of prophylactics and of warnings would be joined into one rule of persuasively creating the standards of "being healthy". The rule would expand the patterns of successful prophylactics with immediate warnings about the potential threats to health which appear in pathological phenomena. The game should not take the form of an absolute negation, an aggressive attack sowing psychosis in the public. The most current example of such an impact is the present media campaign spreading the fear of serious illness among beef-eaters, due to the "British mad cow disease". The sense of threat is created by the media and reaches the level of international-range psychosis. An example of well-done campaigns are programs against smoking tobacco, which have good results in reducing smoking in the United States and Europe mostly among young people.

It must be mentioned that during the transformations of social systems the media, while adapting to new realities, often break the laws with impunity. They cynically take advantage of the peculiar public oversensitivity connected to "the value of health" in those periods. They consciously commit the "iatrogenic fault" of offering the frightened and increasingly poor public a whole series of "miracles and supernatural solutions", every one of which would be an absolute remedy for all illnesses.

Because health is one of the most important and salient values of human existence, one can adopt Machiavelli's hypothesis concerning the desired relationship and mutual impact between the media and health: "The end justifies the means". So, every concept of media which would sufficiently care for public health should be allowed. It should also be remembered that homogenous societies, especially during periods of peril, are vulnerable to media influences according to the "bullet theory", and pluralistic societies, being diverse as far as their systems of accepted norms and values go, are prone to vulnerability toward effective agitation according to the theory of "multi-level flows of information" and the theory of "a message harmonised with cultural and social norms". Finally, one can also claim that in times of peril, due to the "full democracy of death", we all are "homogenous". Consequently, the "bullet theory" seems to be the most successful one, despite still being dangerous and shocking to many people. Hence, one should keep in mind, on the other hand, the main principle of medical treatment: *primum non nocere.*

Swarzdez

REFERENCES

J. Aleksandrowicz, *Sumienie ekologiczne* (Ecological Consciousness), Warszawa, 1988.

J. Bejnardowicz, M. Sokołowska, "Socjologia a pojęcie stanu zdrowia" (Sociology and the Notion of the State of Health), in: M. Sokołowska, J. Hołówka, A. Ostrowski (eds.), *Socjologia a zdrowie* (Sociology and Health), Warszawa, 1976, pp. 417–422.

W. Dieckmann, *Sprache in der Politik. Einführung in die Pragmatik und Semantik der politischen Sprache*, Heidelberg, 1969.

M. L. De Fleur, *Theories of Mass Communication*, New York, 1970.

I. S. Fiut, "Idea bioetyki" (The Idea of Bioethics), in: P. Dudkiewicz (ed.), *Ochrona środowiska w świetle teorii wartości* (Environmental Protection in Light of the Theory of Values), Kraków, 1992, pp. 58–60.

I. S. Fiut, *Filozofia ewolucyjna Konrada Zachariasa Lorenza. Studium problemowe i historyczne* (Evolutionary Philosophy of Konrad Zacharias Lorenz. The Problem and Historical Monograph), Kraków, 1994.

J. Grzybczak, "Czy oddziaływanie mediów jest skuteczne? Stare i nowe teorie" (Are the Media Effective? Old and New Concepts), *Zeszyty Prasoznawcze*, 1995, No. 3–4, pp. 17–39.

E. Katz, "Mass Communication Research and Study of Popular Culture", *Public Opinion Quarterly*, 1957, No. 21, pp. 61–78.

J. T. Klapper, *The Effects of Mass Communication*, New York, 1960.

H. D. Lasswell, N. Leites, and associates, *Language of Politics: Studies in Qualitative Semantics*, revised edition, Cambridge, 1996.

N. Lin, *The Study of Human Communication*, Indianapolis, 1971.

K. Lorenz, "Ecce homo", in: K. Lorenz, *Tak zwane zlo (So-called Evil)*, Warszawa, 1975, pp. 315–346.

N. Maccoby, J. W. Fraquahar, "Communication for Health: Unsealing Heart Disease", *Journal of Communication*, 1975, No. 25, pp. 114–126.

E. Noelle-Neumann, "Return to the Concept of Powerful Mass Media", in: H. Eguchi, K. Sata (eds.): *Studies of Broadcasting: An International Annual of Broadcasting Science*, Tokyo, 1973, pp. 67–112.

W. Pisarek, "Współczesne słowa sztandarowe w Polsce i ich publiczność" (Modern Flag Words in Poland and Their Public), *Zeszyty Prasoznawcze*, 1992, No. 1–2, pp. 16–39.

H. Siwek, "Nastawienie Polaków wobec rzeczywistości społecznej. Raport z badań ankietowych – listopad 1984 rok. Raport OBP" (Attitude of Poles toward the Social Reality. Report from a survey), Kraków, 1985. (Material mimeographed in Archives of OBP.)

E. Roger, F. Shoemaker, *Communication of Innovations*, New York, 1971.

H. J. Schneider, *Zysk z przestepstwa. Środki masowego przekazu a zjawiska kryminalne* (Profit from a Crime. Mass Media and Criminal Cases), Warszawa, 1992.

K. Szewczyk, *Etyka i deontologia lekarska* (Medical Ethics and Deontology), Kraków, 1994, pp. 127–137.

L. M. Walters, "Dobro, zło i media: efekty masowego komunikowania" (Good, Bad and the Media: The Effects of Mass Communication), *Zeszyty Prasoznawcze*, 1995, No. 3–4, pp. 40–49.

C. F. von Weizsäcker, "Modele zdrowia i choroby, dobra i zła, prawdy i fałszu" (The Models of Health and Illness, Good and Bad, Truth and Falsity), in: C. F. von Weizsaëcker, *Jedność przyrody* (The Unity of Nature), Warszawa, 1978, pp. 325–402.

D. Zalewska, "Kulturowy kontekst funkcjonowania zdrowia jako wartości" (Cultural Context of Health Functioning as a Value), in: *Socjologia zdrowia i medycyny. Konwersatorium 1986–88. Materiały* (Sociology of Health and Medicine. A Workshop), Warszawa, 1989, pp. 105–133.

PART IV

THE POLISH SCHOOL OF MEDICINE

ILANA LÖWY

THE POLISH SCHOOL OF PHILOSOPHY OF MEDICINE: LESSONS FROM THE PAST

This meeting celebrates 75 years since the creation of the first chair of "philosophy and history of medicine". The chair, created in Kraków in 1920 for professor Władysław Szumowski, was a double first: the first in the world to explicitly refer to a "philosophy of medicine", and the first to link philosophical and historical studies of medicine. Two thinkers were directly involved in this important innovation: Władysław Szumowski and Adam Wrzosek. Both were medical doctors with broad humanistic interests, and both were interested in promoting interactions between the humanities and medical practice. Władysław Szumowski (1875–1954), born in Warsaw, studied medicine at St. Petersburg, then at Warsaw.[1] He aspired to become a medical researcher, and his first scientific work was a study on the "Culture of tuberculosis bacilli in protein-free media". After obtaining his medical diploma in 1899, he spent two years in Fribourg where he specialized in bacteriology (with Maurice Arthus), then went to Heidelberg and worked in a physiology laboratory (under Albrecht Kossel). His promising research career was, however, cut short by tuberculosis. Diagnosed with this disease in 1902, he returned to Poland and spent four years in a sanitarium in Zakopane. In 1903, Szumowski, who wanted to find an occupation compatible with his health problems, started to study philosophy and history at Lwów University, under Twardowski and Finkel. He wrote a doctoral dissertation in philosophy on "Descartes and Malebranche as precursors of the theory of feelings of Karol Lange", and a historical study on medicine in Galicia (the Kraków region) in the late eighteenth century. In 1907 he obtained both a Ph.D. in philosophy (for the first work), and a certificate in the history of medicine (for the second). From 1907 to 1917, Szumowski taught the history of medicine at the Medical School of Lwów University. In parallel, he did not abandon the practice of medicine. He was at first successively employed by several clinics, and from 1912 on he held a series of jobs in public health.

During his philosophical and historical studies, Szumowski became interested in the work of Polish philosophers of medicine of the previous two generations: Chałubinski (1820–1889), Biegański (1857–1917) and Biernacki (1866–1911). In 1917 he wrote an article on "The Polish

A.-T. Tymieniecka and Z. Zalewski (eds.), Analecta Husserliana LXIV, 295–308.

philosophico-medical school".[2] This article proposes that the medical school's curriculum should include philosophy, and that the two branches of humanistic studies of medicine – history and philosophy – should be gathered into a single subject matter. It is true, Szumowski explains, that such a subject matter does not exist in other countries, but Poland can and should make original contributions in many fields, including the education of physicians. In 1919, with the creation of the Polish State, Szumowski was appointed director of the publication committee of the Health Ministry. He then published an article – "On a subject in medical studies called the history and philosophy of medicine" – in which he developed his previous ideas about the importance of philosophical education for doctors, and about the need to combine philosophical and historical points of view.[3]

The monograph was enthusiastically received by Szumowski's colleague Adam Wrzosek, who was at that time the head of the Department of Higher Education at the Public Instruction Ministry. Wrzosek (1875–1965) had himself been active for years in the field of the history and philosophy of medicine. He studied medicine in Kiev, then in Berlin, and from 1906 on, he taught pathology at the Jagiellonian University of Kraków, specializing in physical anthropology.[4] In parallel, he became interested in humanities in medicine, and from 1910 on added lectures on the history of medicine to other topics he taught at Kraków. Wrzosek published (in 1910) a biography of the Polish doctor Jedrzej Śniadecki, and numerous articles on historical and philosophical topics. In 1918 he published the monograph *On the Necessity of Philosophical Education for Physicians*, which shared some of Szumowski's preoccupations. In this monograph Wrzosek explained that Poles do not need to look for models abroad, and that they should draw on their own philosophico-medical tradition in order to introduce innovative methods to the teaching of medicine.[5] Wrzosek's main contribution was, however, an organizational one; he played a central role in the institutionalization of the history and philosophy of medicine in Poland. Besides his crucial contribution to the establishment of chairs of "history and philosophy of medicine", he founded (in 1924) the journal *Archives of History and Philosophy of Medicine*, which centralized the activity in that field in the inter-war period.

A decree by the Ministry of Public Instruction of November 8, 1920, established five chairs of "history and philosophy of medicine", one in each Polish medical faculty (Kraków, Warsaw, Lwów, Vilnius, and Poznan). The first appointment was that of Szumowski to the Kraków chair (March 1, 1920).[6] He held this chair from 1920 to 1939, and actively developed a method for teaching the history and philosophy of medicine which incorporated the specific Polish tradition in this domain. In 1920 Szumowski founded in Kraków an Institute for the History of Medicine. This institute extended the Museum of the History of Medicine, affiliated with the chair of the History of Medicine of Jagiellonian University created before the war by Wrzosek, and transformed it into an active research center. Szumowski himself was a diligent researcher, specializing in the history of Polish medicine. His main efforts were, however, dedicated to teaching, and his magnum opus was the monumental *History of Medicine from a Philosophical Point of View*, whose three volumes appeared in 1930, 1932 and 1935. The book was based on solid erudition, and had clearly defined didactic goals. One of these goals, Szumowski explained, was – to use an anachronistic term – to provide "role models" for medical students. In his words, "the history of science and history of medicine deal with exceptional individuals. Their lives, their joys and sorrows, have a decisive educational importance for young generations. Physicians have, and should have, a cult of science and great men of science".[7] Such a cult of "great men of science" is reflected e.g. in Szumowski's portrait of Pasteur, whom he describes as "one of the greatest scientists and discoverers in the history of humanity, the immortal creator of bacteriology (...) who was the best possible son, brother, friend, husband and father. (...) Simple, unpretentious and full of sympathy for others, he did not understand irony, despised ridiculing people, and believed that skepticism and a sarcastic attitude were highly destructive forms of human activity".[8]

The uncritical "cult of science" and "cult of great scientists" has disappeared today. It is difficult, however, to blame Szumowski for not sharing our post-modern sensitivities, and our suspicion of "big stories" and "big men", and all the more so if we remember that his hagiographic treatment of Pasteur faithfully reflected the cult of that scientist in France in the early twentieth century.[9] The strong opposition to materialism and positivism expressed in Szumowski's book, is somehow more disturbing. For example, he praises Pasteur for his distrust of positivism, which, according to Szumowski, reflected Pasteur's conviction that the positivists fail to understand the most important notion of all – the infinite.[10] The attempt to mobilize Pasteur against materialism reflects Szumowski's overall philosophical stance, characterized by an opposition to reductionist trends in medicine, and by strong support for "holistic medicine". "Holistic medicine" as we understand it now, that is, the attention to a patient as a complete human being rather than as the "stomach in room number 10" or "the "interesting leukemia case", is usually seen as a positive development in healing. Such "holistic medicine" is often presented as a recent innovation and as a reaction to the recent hyper-technicization and hyper-fractionating of the care of patients after the Second World War. This perception is, however, incorrect. "Holistic medicine" (in France usually called "neo-Hippocratic medicine") was a major trend in the inter-war period.[11] It had numerous branches, some of which were associated with liberal political views, and others with conservative-Catholic ones.

Szumowski expressed affinity with the conservative holistic trends, linked by him to a strong rejection of "materialism" in medicine. In a programmatic article on "The philosophy of medicine as a subject matter of university studies", published in 1920, he explained that the philosophy of medicine is a highly efficient antidote for materialism, a very dangerous doctrine for a doctor that may have disastrous social effects such as the encouragement of hedonism and amoral sexual conduct. He also stressed that for him Władysław Biegański was an exemplary philosopher of medicine because he grounded his philosophical thought in the search for moral virtues.[12] Conservative political or religious views need not be *per se* problematic in a physician. I certainly do not propose that liberal physicians are in some undefinable way "better" professionals than conservative ones, and one may argue that religious faith can increase doctors' commitment to helping suffering persons. In the climate of the 1930's, however, some associations of ideas carried a specific weight. For example, in his review of the philosophy of medicine, Szumowski affirmed that "in 1935, Alexis Carrel, physician, researcher, Nobel prize laureate, wrote a seminal book, *Man the Unknown*. This book, which went through several editions and translations, attributes to medicine a key role in society. It is only necessary, the book explains, to impregnate medicine with larger values, and with a philosophy which takes into account not only the body but also the soul".[13]

Carrel, a conservative Catholic, did indeed center his book on an attempt to combine the newest biological knowledge with spiritual values and ethical considerations. But what values? Carrel wanted to promote the regeneration of the Western countries threatened by the rapid reproduction rate of African and Asian races such as Arabs, Indians and Russians.[14] He viewed democracy – for him, a system which defends the weak and prevents the development of elites – as a sad mistake which luckily was in the process of being corrected through the concrete experience of nations.[15] He proposed that the social classes should become biological ones.[16] And he had an original solution for the problem of the increase in criminality: the rod and the hospital for the less dangerous criminals, and "as to the others, those who have killed, who have robbed with arms, who have kidnapped children, who have stolen from the poor, who seriously have misled the public, an euthanasia establishment, equipped with an appropriate gas, will allow us to eliminate them in a human and economical way. And shouldn't the same treatment be applied to madmen who have committed criminal acts?"[17]

Szumowski's praise for Carrel's book did not mean, of course, that he endorsed similar ideas. This book, I hasten to add, was reissued in France as late as 1962 without any comments or a historical introduction. Carrel remained until very recently a "famous French scientist". Only in the last three years has his name been removed from streets and institutions, thanks to the pressure of anti-racist activists. (Not surprisingly, Carrel has become one of the heroes of the ultra-nationalist movement of Jean Marie Le Pen.) The new criticism of Carrel is also related to the recent and very detailed reevaluation, by the French, of the Vichy government's heritage. (Not surprisingly, again, Carrel enthusiastically supported Pétain, and became one of the official scientists of the Vichy regime.) Still, Carrel's fame in post-war France was based (publicly at least) on his concrete scientific achievements as a pioneer of vascular surgery and tissue culture, not on his "philosophy which takes into account not only the body but also the soul". The uneasiness created by the affinity between some of Szumowski's statements and views propagated by nationalistic-conservative political forces active in the inter-war period is not so easily dispelled. It is accentuated by selected writings of his colleague, the true founder of the Kraków chair, Adam Wrzosek. Wrzosek shared Szumowski's anti-materialist standpoint. He combined it with a belief in the superiority of the Christian outlook. Thus, in a 1927 article, Wrzosek explained that "there was not a single eminent physician in the history of Poland who was an atheist, pantheist or materialist, because all our great physicians, like all our great political leaders, poets and thinkers, were impregnated with deep religiosity in the spirit of the Christian faith".[18] Wrzosek also believed that racial characteristics influence intellectual achievements. In the inauguration talk "Introductory conference on the history of medicine" delivered at the Jagiellonian University (Kraków), in November, 1910, Wrzosek explained that "some nations, for example the Chinese, have no science, because they do not possess even a tiny crumb of scientific creativity. The typical traits of every nation are reflected in its scientific output. The historian of science should therefore take into account the contribution of the racial factor to the creativity of each nation".[19] Later Wrzosek tried to find empirical confirmation for his theories. A 1931 article, "On the relationship between anthropological measurements, racial types and intellectual abilities", is based on an investigation which attempted to find correlations between racial types (based on anthropomorphic measurements) and intelligence (as measured by an intelligence test developed by Prof. Bayle). Here 173 students (all of Polish nationality and Catholic faith, to ensure the uniformity of the sample) were tested, and the results of these tests demonstrated that students of the Nordic type were more intelligent than students of the Oriental type.[20]

Should we then simply classify Wrzosek and Szumowski as holders of a backward ideology and "politically incorrect" views, even if we mitigate our condemnation by the recognition of the fact that their opinions were considered acceptable and non-exceptional before the Second World War? And what about their important contribution to the diffusion of the thought of the Polish School of Philosophy of Medicine, their efforts to provide a humanistic education for doctors, and their innovative linking of philosophy and the history of medicine? Should these achievements be dismissed? One may be tempted to label the parts of Wrzosek's and Szumowski's thought which make us uneasy as a "mistake" and as "just reflecting their times", and separate them from what we can see now as the important kernel of their heritage. But how should such a separation between "mistaken" and "essential" elements of a work be made, and who is entitled to make it?

To avoid the – perhaps justified – suspicion of priggishness (it is easy enough today to denounce racial prejudice and religious bigotry), I would like to mention briefly another example of uneasy associations, one that concerns me directly as a social historian of medicine. The preface to the 1961 edition of Szumowski's *History of Medicine*, written by Prof. Dr. Bolesław Skarzýński, praises Szumowski's historical erudition but criticizes the ideological background of his work. The history of medicine, Skarzýński explains, "should go beyond the mere description of medical facts and situate historical events in medicine in a larger political, cultural and economic context". The majority of my colleagues will probably agree with this statement, but will be less enthusiastic about the following one: "The history of medicine should develop a synthetic approach, grounded in the scientific principles of materialist dialectics", and should learn from developments in this area in the Soviet Union.[21] I am well aware of the fact that Prof. Skarzýński's text might have been, and probably was, a ritual formula necessary to obtain permission to publish an "ideologically suspect" study. But even if that was the case, I find disturbing the close association between the (for me, correct) call to situate historical studies of medicine in a broader social, cultural and economic context, and the (for me, incorrect) call to apply materialist dialectics, Soviet-style, to these studies. It is disturbing mainly because we know that in other circumstances the association of an interest in the social history of sciences and of enthusiasm for the achievements of the USSR was not limited to standard formulas and empty declarations. I am a member of an international association of social historians of medicine, named "The Sigerist Circle" after one of the founders of this field, Henry Sigerist. In 1937, Sigerist, then a professor of the history of medicine at Johns Hopkins University, published a book on Soviet medicine. In its introduction, Sigerist explained that he "approached the problem of soviet medicine as a historian, studying and analyzing it as a historical phenomenon". The result of this scientific effort was a categorical affirmation that the Soviet regime was the most democratic society in the world, that the Soviet Union was the only country which has a truly free press, and that it had created a "new civilization with boundless possibilities of development".[22] Skarzýński's text thus forces us to reflect upon a subject which some social historians of science may find uncomfortable: the problematic past of selected areas of "progressive historical thought".

With hindsight, we claim to know what the "right" and "wrong" ideas are. We know that the humanistic approach to medicine is good and that racism is bad, that social history is good scholarship and that the Soviet version of materialist dialectics is bad. We may also try to separate the wheat and the chaff, the good ideas from the impure contaminants. In "their time", however, the separation between what we now see as right and wrong was not so obvious. Today we are quick to spot the irreducible contradiction between Carrel's programmatic affirmation that his aim was to develop a science of man taking into account the wholeness and the uniqueness of the individual, and his recommendation to develop euthanasic establishments for a rapid and economical extermination of undesirable persons, but this contradiction was much less visible in the 1930s.[23] Perhaps the most troubling question, then, is how we can be sure that "we" are so different from "them", and that the present discourse, say on "humanistic medicine", is free from some unwanted "influences of the period" that are hidden from us now but that will become obvious from a historical perspective. The last question is directly borrowed from a recent article by the historian of medicine Anne Harrington, who encountered similar questions in her studies of the holistic trends in German medicine, and stumbled over the case of Victor von Weizsäcker.[24] Weizsäcker was a powerful spokesman for humanistic medicine in post-war Germany, and a physician deeply involved in the anthropological approach to diseases. He developed his holistic perspective on medicine in the pre-war period, and continued to advocate it until his death in 1957. Weizsäcker is often presented as an emblematic figure of German medical humanism. However, during a conference in 1986 which celebrated Weizsäcker's one hundredth birthday, one of the participants declared that new documents, found in a military archive in Katowice, shed a suspicious light on Weizsäcker's activities during the Second World War. According to these documents, Weizsäcker, who was the head of the Neurological Clinics and Research Institute at the University of Breslau (Wrocław) between 1940 and 1945, asked for, and received for study, preparations of brains of "asocial and handicapped children" from the Lubliniec Psychiatric Clinics for Youth, "eliminated" in the framework of the Nazi euthanasia program. Weizsäcker's daughter then asked what these documents had to do with her father's work, but the conference chairman rapidly dismissed her question and the whole topic by saying that it should be clarified elsewhere. Weizsäcker's daughter's question, Harrington proposes, concerns us, too. It is not only about the relationships between humanistic theories and doubtful acts, but also about the challenge to look at the past not from the position of moral outrage but with the "eyes of identity". If we do so, we may discover that "past images and phantasms of goodness could have seduced us too". We may then lose our certainty that "our current commitments to the good causes and ideals around which we have built our professional lives are allowing us to see as clearly as we should".[25]

Anne Harrington, I propose, is right when she insists on the ambiguous uses of some visions of salvation and reform, and warns us to beware of professional smugness and self-righteous feelings when evaluating historical evidence. I find her attitude more questionable when (in the above-mentioned article) she leaves the question of the evaluation of Weizsäcker's activities open, and limits her discussion of his case to abstract reflections on evil which may lie hidden even in the most angelic language. In Weizsäcker's case the evil, if it was there, was not related to humanistic or other language, but to barbaric practices. Her readers, at least the non-expert ones, would like to know more about Weizsäcker's history. Harrington's article was published in 1995, thus nine years after the documents allegating Weizsäcker's suspicious war activities were first mentioned. What has happened since? Was the information that Weizsäcker not only used "research materials" from euthanized children, but knew about the origins of this "material", confirmed? If it was, then its significance should be evaluated together with all the other things we know about Weizsäcker, especially about his behavior during the Third Reich years (what were his political opinions? Did he actively or passively oppose the regime? Did he assist its victims?) and about his work as the head of the neurological Clinics at the University of Breslau. (Was he associated with other doubtful activities? Was it possible for him to have had an influence over the euthanasia program in the Lubliniec Clinics?) On the other hand, if the question of Weizsäcker's use of brains of assassinated children remains open – and it is quite possible that no certainty can be achieved even with the most careful historical investigation – we still are left with a problem of evaluation of the activities of an individual doctor in a specific historical context. Unambiguous moral judgments of past events are difficult, among other things because we are often facing incomplete data. This difficulty does not mean, however, that the absence of certainty about the past, and the existence of unavoidable flaws in our well-meaning discourse, should prevent us from making such judgments. Elsewhere, Anne Harrington does not hesitate to provide "contextualized" appraisals of past activities. She provides a sympathetic account of the work of the humanistic German psychiatrist Kurt Goldstein,[26] and gives a chilling description of the growing and testing of herbal medicines at the Dachau concentration camp – an enterprise organized by holistic medicine adepts in the SS, and conducted with a rapidly rotating population of prisoners.[27]

We may thus agree that the evaluation of a given action should not be dissociated from the context of this action. Even a racist discourse may be seen as relatively harmless if pronounced in a society which has no contacts with ethnic minorities, or one in which there is a strong consensus against such views. An agreement on the desirability of "contextualization" does not, however, solve the question of how to study "contextualized" actions. Here, I propose, we can benefit from what I believe is the most important innovation made by the Polish School of Philosophy of Medicine – the tradition of "philosophizing at the bedside", that is, of theoretical reflections on situated practices. The Polish philosophers stressed this unique orientation of their school of thought. Biegański highlighted the difference between the Western philosophy of medicine, interested in theoretical and scientific aspects of medicine, and the Polish philosophy of medicine, interested in its practical tasks. Biernacki explained that the Polish theoreticians of medicine make explicit "the idea hidden in the everyday activity and behavior of the physician", while Szumowski proposed that in Poland the philosophy of medicine was "the child of the clinics". But perhaps the most efficient description of the particular bent of the Polish School of Philosophy of Medicine came from the last and most famous thinker associated with this school, Ludwik Fleck (1896-1961), today viewed as a pioneer of social studies of science. In his article "On the Crisis of 'Reality'", written in 1929,

Fleck proposed to look carefully at what the scientists do in the laboratory, because otherwise "one mistakes the natural sciences as they are, for the natural sciences as they ought to be, or rather as one would like them to be".[28] Fleck's highly original methodological proposition, I have argued elsewhere, has its roots in the practice-oriented reflections of Polish philosophers of medicine.[29] This tradition can still teach us how to examine concrete actions in concrete settings.

Two examples, one taken from the work of Zygmunt Kramsztyk (1848-1920), and another from Fleck's own studies, show us how these thinkers grounded their theoretical reflections in a careful observation of doctors' practices. Kramsztyk was concerned by the ways in which physicians use the concept of "rational therapy". "Years ago", he tells us, "I was asked to consult with two colleagues in order to give advice on a tabetic patient. The most respectable of us said: 'The nervous system of the patient is ruined – it needs stimulation. But since we cannot penetrate into its very center, we will have to send an impulse through peripheral nerves; the skin nerves have to be stimulated. Let us advise a spa treatment at Ciechocinek for the patient'". And so it was decided. I was deeply depressed. If it had been I who advised the patient to go to Ciechocinek, it would have been because a treatment in Ciechocinek seemed to be helpful for other patients. The advice of my learned colleagues sounded, however, like an absolute necessity. (...) Many times I have met colleagues who challenged my inclination for this type of rationalism and finally I have discovered its secret. The secret consists in a kind of mental tunnel. The rational doctor starts his action in the same way as the average mortal. First of all, he goes to the medicine chest, picks up a bottle and vanishes in the tunnel. Then he reappears at the other end with the bottle in hand. Anyone seeing, or rather listening to him, is wondering where the hell the chap got castor oil in the first place? But he just carried it through the tunnel.

"In other words, being in possession of a medicine, the doctor searches for a detail of the case that can be connected in a satisfactory manner with what he already has got in his hand. The rational man travels both ways, from the therapeutic indication to the symptom, and back again. But the first part of the journey is concealed from the world and here is where all the magic originates. For example, in some inflammations a cold compress was usually applied: 'My dear colleague, the blood vessels are dilated. When we apply an ice compress, they will contract and the patient will recover'. How clear and simple! However, a couple of years later this method of treatment has changed. It has been found by that time that a hot poultice yields better results. Hot poultice? This is good too! 'My dear colleague. Hot poultice is indispensable for the patient; the exudate is hard – it must be softened to enhance suppuration'. But sometimes the explanation is not so easy. There are cases in which the disease is so remote from the therapeutic means to be recommended that linking the two becomes a problem. The distance between tabes and Ciechocinek is too long. There is no way to cover it in one jump. Therefore, an intermediate station has to be selected – say, skin stimulation. First the spinal cord is tied to the skin and, well from the skin to Ciechocinek the road is straight enough".[30]

Fleck was interested in the circulation of "scientific facts" between different professional communities and their transformation during the passage from one community (in Fleck's terms, thought collective) to another. One of his examples was the bacteriological diagnosis of diphtheria. An expert bacteriologist discussing a throat swab with another expert, will indicate that many, but not all the bacteria in the studied preparation have the characteristics of Loffler bacillus, the etiologic agent of diphtheria. A description intended for the expert will be something like this: "Numerous bacilli, many of them club-shaped and slightly curved (...) their arrangement is in several places finger- and palisade-shaped, elsewhere irregular (...) Loffler methyl blue: many lacerated bacilli (...) sharply defined colonies, in which bacilli were found mostly typical in their staining characteristics, morphology and arrangement". The expert's conclusion, as formulated for his peers, is that in the view of the origin of the examined material and the morphological and culture characteristics of the bacilli, the diagnostic of Loffler bacilli seems sufficiently well established. For a professional bacteriologist, Fleck explains, such a presentation of a diagnosis is as close to certainty as possible. Everything – the arrangement of the bacilli, the staining and the culture – is in agreement. It is not acceptable, however, to a general practitioner who expects from the expert a firm diagnostic statement, not the opinion that "many" bacteria show "mostly typical" traits. In addition, the practitioner may claim: "I just asked you what this throat swab really contains, and you reply – because it is a throat swab, the conclusion is justified that this is diphtheria. That is being mischievous. I wanted your support but you used me to support yourself". Thus the specialist would formulate a different description for the general practitioner: "The microscopic specimen shows numerous small rods whose shape and position correspond to those of diphtheria bacilli. Cultures grown from them produced typical Loffler bacilli". And for the patient, the description will be even shorter. The doctor will simply state that he has diphtheria.[31]

Kramsztyk and Fleck tell deceptively simple stories. In fact, however, these stories skillfully display multilevel relationships between medical discourses and practices. Their studies indicate that in order to have a "contextualized" understanding of what is going on in a given field (be it medicine or history) one should follow what its practitioners do – or, to be more precise, follow what they do, what they say while they do it, what they claim they do, and what the concrete consequences of their deeds are – and then try to creatively link all these data. Philosophical debates centered on values, and a collective reflection on how things should ideally be done, are undoubtedly necessary and important. But, to follow the thinkers of the Polish School of Philosophy of Medicine, such debates should be accompanied by attempts to develop critical reflections on situated actions. Fleck described science as "an eternal, synthetic rather than analytic, never-ending labor – eternal because it resembles that of a river that is cutting its own bed".[32] The river metaphor is particularly effective because it illuminates the connections between a "fact" and its "context", and the complicated ways in which the past is inscribed in the present. And, one should remember, rivers more often cut their bed in muddy soil than in a solid rock. When trying to understand the past, we are often obliged to try to make sense of partially obliterated traces, turbid paths, confused signs, and obscure configurations. An effort is necessary to understand past and present actions in their context, and while making this effort we should, as Anne Harrington forcibly reminds us, beware of complacent attitudes and be prepared for uneasy confrontations. Kramsztyk would have agreed: for him, "history prevents arrogance and excessive confidence in present ideas. It teaches us modesty, because it helps us to perceive the instability of every single opinion".[33]

Hôpital des Enfants Malades, Paris

NOTES

[1] On Szumowski's career, see Władysław Szumowski, "La Philosophie de la médecine, son histoire, son essence, sa dénomination et sa définition", paper presented at the 5th International Congress of the History of Sciences, Lausanne, October, 1947. *Archives Internationales d'Histoire des Sciences*, 1949, **2**, 1097–1139, pp. 1112–1115. Tadeusz Bilikiewicz, "From the Editorial Committee", in: Władysław Szumowski, *Historia Medycyny* (new edition, reviewed and corrected by an editorial committee headed by Tadeusz Bilikiewicz). Warsaw: Państwowy Zakład Wydawnictw Lekarskich, 1961, pp. x-xiv.

[2] Władysław Szumowski, "Polska szkoła filozoficzno-lekarska", *Polski Miesięcznik Lekarski*, 1917, 5–6.

3 Władysław Szumowski, "O przedmiocie studiów lekarskich zwanym historia i filozofia medycyny", *Gazeta Lekarska*, 1919, **11**.

4 Teresa Ostrowska, "Profesor Adam Wrzosek i jego dorobek naukowy", *Wiadomości Lekarskie*, 1977, **30** (1), 63–66.

5 Adam Wrzosek, *O potrzebie studiów filozoficznych dla lekarzy*. Kraków: Drukarnia Uniwersytetu Jagiellońskiego, 1918.

6 The Warsaw chair was attributed to Franciszek Giedroyć, who delegated the teaching of philosophy to Henryk Nusbaum; the Vilnius chair was attributed to Stanisław Trzebiński, the Poznań chair to Adam Wrzosek, while the Lwów chair remained empty until 1931, when it was attributed to Witold Ziębicki.

7 W. Szumowski, *Historia Medycyny*, op. cit., p. 378.

8 Ibid., pp. 344–345.

9 On Pasteur's hagiography, see: Daniel Raichvag, *L'Empire des microbes*. Paris: Gallimard (Collection Découvertes, Sciences), 1995. Bruno Latour, *Pasteur. Une science, un style, un siècle*. Paris: Librairie Académique Perin Institut Pasteur, 1994.

10 W. Szumowski, *Historia Medycyny*, op. cit., p. 345.

11 Proceedings of the meeting, "Holistic Medicine", McGill University, Montreal, May 3–5, 1995.

12 Władysław Szumowski, "Filozofia medycyny jako przedmiot uniwersytecki", *Przegląd Filozoficzny*, 1920, 23.

13 W. Szumowski, "La Philosophie de la médecine, son histoire, son essence, sa dénomination et sa définition", op. cit., on p. 1126. A review of the 1948 edition of Szumowski's book *The Philosophy of Medicine*, stressed the interest Szumowski had for Carrel's views. Arpad Herczeg, "Compte rendu du livre de Władysław Szumowski, *Filozofia Medycyny*". Kraków: Gebethner & Wolf, 1948. *Archives Internationales de l'Histoire des Sciences*, 1950, 29, 220–222.

14 Alexis Carrel, *L'Homme cet inconnu*. Paris: Plon, 1962 (1935) p. 18. Carrel's book was reissued in a pocket edition in 1962.

15 Ibid., p. 374.

16 Ibid., p. 409.

17 Ibid., p. 435–436. On the history of the application of this idea in Nazi Germany from 1939 on, see Robert N. Proctor, *Racial Hygiene: Medicine under the Nazis*. Cambridge, Mass: Harvard University Press, 1988. Adam Wrzosek, *Wykład wstępny z historii medycyny* (November 3, 1910, Jagiellonian University, Kraków). Lwów: Drukarnia Piller-Neumann, 1911.

18 Adam Wrzosek, "Religijność naszych znakomitych lekarzy w XIX wieku", *Archiwum Historii i Filozofii Medycyny*, 1927, **7**, 159–167.

19 Adam Wrzosek, *Wykład wstępny z historii medycyny* (November 3, 1910, Jagiellonian University, Kraków). Lwów: Drukarnia Piller-Neumann, 1911.

20 Adam Wrzosek, "O stosunku niektórych pomiarów antropologicznych i typów rasowych do sprawności umysłowej", *Przegląd Antropologiczny*, 1931, **5**, 1–15.

21 Bolesław Skarzyński, "Słowo wstępne", in: Szumowski, *Historia medycyny*, op. cit., pp. m–lx, on p. viii.

22 Henry E. Sigerist, *Socialized Medicine in the Soviet Union*. London: Victor Gollancz, 1937, pp. 19–22, 322–326.

23 On Carrel's goals, see Carrel, *L'Homme cet inconnu*, op. cit., p. 344.

24 Anne Harrington, "Unmasking Suffering's Masks: Reflections on Old and New Memories of Nazi Medicine", *Dedalus*, Winter 1996, 125 (1), 181–206.

25 Ibid., pp. 199–200.

[26] Anne Harrington, "Kurt Goldstein's Neurology of Healing and Wholeness: A Weimar Story", paper given at the conference "Holistic Medicine", op. cit.

[27] Harrington, "Unmasking Suffering's Masks", op. cit., p. 194. This is clearly an extreme case: not every "context" so obviously shapes every action within it as the Dachau concentration camp did.

[28] Ludwik Fleck, "On the Crisis of 'Reality'", (trans. H. G. Shalit and Y. Elkana), in Robert S. Cohen and Thomas Schnelle (eds.), *Cognition and Fact: Materials on Ludwik Fleck*. Dordrecht: D. Reidel, 1986, pp. 47–58, on p. 51. This article was first published in German in 1929.

[29] Ilana Löwy, *The Polish School of Philosophy of Medicine: From Tytus Chałubiński (1820–1889) to Ludwik Fleck (1896–1961)*. Dordrecht: Kluwer, 1990, pp. 215–228.

[30] Zygmunt Kramsztyk, "Racyonalne leczenie", in: Z. Kramsztyk, *Szkice krytyczne z zakresu medycyny*. Warsaw: E. Wende, pp. 189–196.

[31] Ludwik Fleck, *Genesis and Development of a Scientific Fact* (trans. F. Bradley and T. Trenn). Chicago: University of Chicago Press, 1979 (1935), pp. 113–114.

[32] Fleck, "On the Crisis of 'Reality'", op. cit., p. 54. The comparison between science and a river was first proposed by Zygmunt Kramsztyk, "O znaczeniu wiedzy historycznej", *Krytyka Lekarska*, 1899, **3** (9), 253–255.

[33] Kramsztyk, "O znaczeniu wiedzy historycznej", op. cit. p. 255.

WŁADYSŁAW SZUMOWSKI

WŁADYSŁAW SZUMOWSKI: FORERUNNER OF CONTEMPORARY PHILOSOPHY OF MEDICINE

In the life of Professor Władysław Szumowski, the founder of the Chair and Department of the History and Philosophy of Medicine, four turning points can be distinguished, each of which had a decisive influence upon his academic career. The first occurred when he had to interrupt his experimental work on tuberculosis – which he had been carrying out after completing his medical studies at the Medical Faculty of the University of Warsaw – when he himself succumbed to tuberculosis infection. The second turning point took place when he followed the advice of Professor Gluziński, Director of the Clinic for Internal Medicine at the University of Lwów, and began to study philosophy under Professor Twardowski. The third came with the outbreak of World War II and his resulting imprisonment along with other professors of the Jagiellonian University by the Nazi-Germans, which interrupted his research and studies for at least six years. The fourth came in 1947 when, after his retirement during the new communist regime, he was forced to witness the withdrawal of the history of medicine and medical philosophy courses from the study programs offered by post-war Polish medical faculties.

Looking back over the 75 years that have passed since Professor Szumowski founded the first Chair of the History and Philosophy of Medicine after presenting his proposal to the Medical Faculty of the Jagiellonian University, a number of questions related to this event emerge. One question is to what extent the foundation of the chair was in accordance with the demands of medical education at that time. Another question, related to the first, is, what arguments survive to the present day justifying the foundation of the chair?

In attempting to partially answer these questions, we may refer to one of Professor Szumowski's own papers, published to commemorate Professor Twardowski, who held the Chair of Philosophy at the University of Lwów at that time and who was Szumowski's teacher. In this paper, entitled "The Philosophy of Medicine as a Subject of University Teaching" (1921), he wrote:

> the Philosophy of Medicine is a science which considers medicine in its entirety, examines its position in the humanities, society, state and medical schools, provides a glance over the history of medicine as a whole, presents general problems related to the philosophy of biology, analyses

A.-T. Tymieniecka and Z. Zalewski (eds.), Analecta Husserliana LXIV, 309–315.

the methodology of medical thinking, enumerates and explains mistakes made in medicine, takes from philosophy and metaphysics such knowledge and views which are of significance and value for medicine in general, defines the basic values underlying medicine and creates the general basis for medical ethics.

As may be concluded from this statement, Szumowski saddled the Philosophy of Medicine with a number of different tasks and responsibilities. Have any of them lost their appeal and validity in our own present-day conditions? It seems that almost all of them have maintained their importance and are still the subject of study, research and concern in our own times. Some of them have even gained in importance, including questions related to logic, the causal relationships between various biological phenomena and, finally, mind and body problems.

At the time of the chair's foundation, contemporary medical thinking on the position of sick people in society was a very important issue for Szumowski. During this period, his understanding of the patient's position was based primarily on rapid developments in medical science and knowledge. The roots of his thinking, however, were anchored in achievements made in the nineteenth century. Szumowski was convinced that the sick were primarily a subject of concern for the family, the closest members of the community, as well as for those voluntarily or professionally engaged in taking care of them. He recognised as necessary the many efforts being undertaken to discover appropriate ways of treating the sick. Today we can observe rapid evolution in health care. As the community has become more highly developed and organised, and as the ways of treating the sick have become and are continuing to be more sophisticated, so the criteria through which health care is realised are also changing more rapidly.

At the time when Szumowski was carrying out his work, the Hippocratic code was the principal code accepted by the majority of the medical community. The ideas encoded in the Hippocratic oath were the result of long-standing observations and thinking on the physician-patient relationship which had greatly influenced the nature of medical work. In modern times scientific achievements have entered into medicine on a much wider scale compared with previous periods in its history. As science has progressed, so has the accumulation of data which can be applied in the medical treatment of a patient. While Szumowski was a witness to such changes, he also attached importance to the past. As a result, he entitled his monograph, "The History of Medicine Philosophically Assumed or Interpreted". In turning toward the past, he appreciated, among other developments, the rationalistic philosophy of Descartes as a break-through event which had an immense impact on various scientific disciplines, among them biology. On the one hand, Cartesian rationalistic philosophy paved the way for a mechanistic concept of the human body and its functions, while on the other, what was probably much more important was that it was a creative force for future scientific achievements contributing significantly to the definition of the criteria of science. One of the first adherents of Cartesian philosophy was La Mettrie, who advanced the mechanistic concept of the human body, presenting his ideas in "L'Homme machine". But from an early period, human thinking very often betrayed a parallelism strongly based on the coexistence of materialism and idealism. The purely materialistic point of view was usually balanced in a more or less pronounced fashion by idealistic views. The same trend occurred in La Mettrie's work. His efforts at explaining his mechanistic views on interpreting the structure and function of the human body were in accordance with the ideas of certain other thinkers on the human body and its existence in health and disease, e.g., von Helmholtz, Sydenham and Stahl, and concorded especially with the views forwarded by scholars based at the university in Montpellier, where Borden and Barthuz were at that time discussing the role of vital force. However, they were strongly attacked by mechanists and materialists. Seeking arguments in favour of medical philosophy, its name and definition, Szumowski quoted Zimmerman, a Swiss physician and the author of *Uber die Erfahrung in der Arzeneikunst*, in which he highlighted the importance of experiments and experience in the medical sciences. This was in the mid-eighteenth century. According to Szumowski, only certain unfavourable conditions during the French Revolution prevented a philosophy of medicine chair from being founded in the newly organised medical faculty of the University of Paris. Two individuals, Cabanis and Thured, were the principle figures actively involved in these efforts. Unfortunately, Szumowski asserted, just as these ideas had appeared like flashes of lightning, so they also suddenly disappeared in a similar fashion. The endeavours of Cabanis were hindered by Berlussan and therefore no lectures on medical philosophy could be given at the university. In the end, Cabanis' life was brought to an end by a stroke, and with his death, the efforts to found a chair of philosophy of medicine eventually came to an end.

The rapid development of the natural sciences in the nineteenth century saw an abundance of works more or less directly related to the subject we understand as the philosophy of medicine. In certain cases, it was even directly called "Philosophie der Medizin", as it was in one of the works authored by Grohmann. Similar studies appeared in other countries, such as France: Bouillard, Auber, Forget; as well as in the U.S.A.: F. Bartlett (*Essay on the Philosophy of Medical Science*, New York), to mention only a few from the long list of authors who were publishing theoretical works at that time on various problems related to medicine.

Among the authors of such contributions were many physicians of Polish descent, including Tytus Chałubiński who created the theoretical basis for medical practice. Chałubiński was the author of a paper called "The method of searching after medical indications" with the subtitle "The planning of treatment and its execution". Chałubiński's ideas were probably based on the work of Lotze, whose writings touched, among other areas, on the utility of logic in medical practice. Seeking arguments justifying the introduction of the philosophy of medicine into medical studies programs, Szumowski ascribed great significance to J. S. Mill's monumental work *The System of Logic*. Among the many works based on Mill's achievements were those of Osterlen, who in 1852 published *Medizinische Logik* (Tubingen), as well as the writings of Bain, whose presentation of logic was in line with Mill's own ideas. These and other studies strengthened Szumowski in his endeavours to teach the philosophy of medicine as a discipline. He also found great support in the significant output made in this area by Polish physicians. Discussions on the theoretical problems of medicine in the Polish Medical Society were very animated during this period. Before Chałubiński did his own work, Dworzaczek, who suffered from blindness, had carried out some studies on the theoretical problems of medicine. His paper, "An introduction to the philosophy of medicine", was presented at a Meeting of the Warsaw Medical Society on 20th January 1856.

Chałubiński's work had its successors in Biernacki, Kramsztyk and, later, Biegański. In considering the achievements of this group, Szumowski found logic to be the most valuable tool of the medical profession, and a knowledge of causal relationships to be the basis of medical thinking.

Szumowski was convinced that mistakes made in diagnosis and treatment would be considerably reduced if physicians were well trained in medical logic. While preparing his Ph.D. thesis as a student under Twardowski, Szumowski lectured on medical logic. After almost 20 years of teaching and lecturing, Szumowski summarised his experiences in *Logic for Medical Students*, published in 1939. This book contains a couple of chapters on causal relationships and inductive thinking which carry the same weight today as they did when they first appeared. Szumowski argued that

> stable and univocal results could always be expected if it were possible to study simple events which remain in direct causal relation to one another. In biology, however, there are no simple events and causal relationships are a product of a number of sequences existing in indirect

relation to one another. It is impossible to dissociate such events and reduce them under observation to a single one because they are all connected in various different ways in the living body.

This type of thinking evolved into a new method for solving medical problems and raised the need for critical evaluation of both the experimental results obtained through this process and their application in human medicine. The importance that the problem of causation held for him is illustrated in the Latin proverb he used in his book: "Vere scire per causas scire". How many of Szumowski's achievements in theoretical medicine have withstood the test of time?

If we compare the titles of the first five chapters of *The Philosophy of Medicine: An Introduction*, by Wulff, Pedersen and Rosenberg, with the papers published by Szumowski, hardly any difference is noticeable with regard to the main issues discussed. Looking at the similarities between Szumowski's work and that of Wulff et al., it seems that today, just as 75 years ago, the issues of greatest importance are: empiricism and rationalism as philosophical problems, empiricism and rationalism as the two current trends in medical thinking, the mechanistic model of the human body, logic as the basic method for solving medical problems, and the principle of causation. The only divergence between the two works is the first chapter of Wulff et al., in which the subject of medical paradigms was discussed, taking into account Kuhn's theory of science. It follows from this comparison that Szumowski had already recognised these themes as the most essential dilemmas in medicine decades ago. Already at a very early stage he had understood the importance of medical logic in medical education. He understood logic not only as a tool of correct thinking and reasoning, but also as a method for solving medical problems. If we compare Szumowski's scheme of causal relationships, which was presented in the form of a sequence of letters used to illustrate the relations between the individual elements of what he called a "causal net", with that given by the aforementioned contemporary authors, no significant differences can be observed between the two approaches. Szumowski, in describing causal relationships in the form of his "causal net", made use of the same symbols that were to be employed by Macki years later. Using this scheme, Szumowski explained the interrelations between the elements of the causal net themselves by paying attention to the conditions under which the process as a whole develops in living organisms. In Szumowski's view a necessary simplification of the system can be observed. To understand the correctness of medical thinking, he introduced the three principles of J. S. Mill in his book: 1) The principle of consistency, 2) the principle of difference, 3) the principle of associated changes. Of the principles and/or methods cited, consistency is the method of thinking which constitutes the logical scheme of observation. The method of difference represents the basic instrument for evaluating experiment results. The method of associated changes has its application in both observation and experiment. All three are basically inductive, using induction as a way of advancing from individual details to generalisations (wholes).

Writing on the same subject 60 years later in *Scientific Reduction and the Essential Incompleteness of all Science*, Karl Popper argued that:

> there are two different ways in which we can look for simplification. They can be briefly termed as philosophical and scientific reduction. The former is characteristic in its endeavour to simplify our view of the surrounding world, the latter enhances the endeavour of daring and testable theories for verification, theories which can gain a powerful basis of their explanation.

Popper was convinced that the latter procedure represented a valuable and workable method that could be used in the biological sciences. Bearing in mind the difference in stature between Szumowski and Popper, I merely wished to say that in the early part of the twentieth century Szumowski was ascribing a value to certain phenomena which years later were to become the main subject of study in modern philosophy. With the development of the modern philosophy of medicine, Szumowski aimed to create a balance between rationalism, realism and empiricism. To the end, Szumowski stressed the particular importance of this area because he was convinced that it had a significant impact on the physician-patient relationship.

The rapidly growing interest in medical ethics and philosophy – an interest that has not been limited to the medical community alone – is certainly linked with the development of science and advances in socio-economic living conditions. Science has created new standards of diagnosis and treatment, pushing into the background the role of experience which, during Szumowski's professional life, had been one of the most important tools of medical practice. Today, rationality has become the chief priority in medical treatment, and all other elements ultimately capable of facilitating accurate diagnosis and treatment have either become secondary factors in modern-day medical practice, or been disregarded completely.

For this reason, it may be useful to recall the words of Professor Jan Miodoński, professor of laryngology, distinguished humanist and a younger colleague of Professor Szumowski's at the Medical Faculty, when he argued:

> I am not able to define Homo Sapiens, but I am convinced that I know in what phase of development the human being can be recognised as man. It will be that phase of development in

which the human creature discloses its ability to feel compassion and charity. It is at this moment that the conditions for the birth of medicine make their appearance. Therefore we may say that medicine is as old as mankind itself. For medicine as we understand it is a subject of interest concerning not only disease but also death, not just physical pain, but also misery and calamity, not simply physical help but also moral support, the basic forms of which exhibit an unusual firmness and stability.

It may be that these principles will survive in the years to come. If so, Szumowski has made his own contribution to this state of affairs.

Kraków

BIBLIOGRAPHY

[1] Szumowski, W. (1920), "Filozofia medycyny jako przedmiot uniwersytecki", Ksiega Pamiątkowa ku czci prof. Twardowskiego [The philosophy of medicine as a university subject, a commemorative work in honour of Professor Twardowski], *Przegląd Filozoficzny,* **43** (1).
[2] Szumowski, W. (1925), "O zwiazku przyczynowym w medycynie" [On the causal relationship in medicine], *Archiwum Historii Medycyny,* **11**: 1–12.
[3] Szumowski, W. (1939), *Logika dla medyków* [Logic for medical students], Kraków: Gebethner i Wolff.
[4] Szumowski, W. (1947), *Dzieje filozofii medycyny. Jej istota, nazwa i definicja* [The history of the philosophy of medicine: its basis, name and definition], Kraków: PAU.
[5] Szumowski, W. (1948), *Filozofia medycyny* [The philosophy of medicine], Kraków: Gebethner i Wolf.

INDEX OF NAMES

Analecta Husserliana

The Yearbook of Phenomenological Research

Editor-in-Chief

Anna-Teresa Tymieniecka

The World Institute for Advanced Phenomenological Research and Learning, Belmont, Massachusetts, U.S.A.

1. Tymieniecka, A-T. (ed.), *Volume 1 of Analecta Husserliana.* 1971 ISBN 90-277-0171-7
2. Tymieniecka, A-T. (ed.), *The Later Husserl and the Idea of Phenomenology.* Idealism – Realism, Historicity and Nature. 1972 ISBN 90-277-0223-3
3. Tymieniecka, A-T. (ed.), *The Phenomenological Realism of the Possible Worlds.* The "A Priori', Activity and Passivity of Consciousness, Phenomenology and Nature. 1974 ISBN 90-277-0426-0
4. Tymieniecka, A-T. (ed.), *Ingardeniana.* A Spectrum of Specialised Studies Establishing the Field of Research. 1976 ISBN 90-277-0628-X
5. Tymieniecka, A-T. (ed.), *The Crisis of Culture.* Steps to Reopen the Phenomenological Investigation of Man. 1976 ISBN 90-277-0632-8
6. Tymieniecka, A-T. (ed.), *The Self and the Other.* The Irreducible Element in Man, Part I. 1977 ISBN 90-277-0759-6
7. Tymieniecka, A-T. (ed.), *The Human Being in Action.* The Irreducible Element in Man, Part II. 1978 ISBN 90-277-0884-3
8. Nitta, Y. and Hirotaka Tatematsu (eds.), *Japanese Phenomenology.* Phenomenology as the Trans-cultural Philosophical Approach. 1979 ISBN 90-277-0924-6
9. Tymieniecka, A-T. (ed.), *The Teleologies in Husserlian Phenomenology.* The Irreducible Element in Man, Part III. 1979 ISBN 90-277-0981-5
10. Wojtyła, K., *The Acting Person.* Translated from Polish by A. Potocki. 1979 ISBN Hb 90-277-0969-6; Pb 90-277-0985-8
11. Ales Bello, A. (ed.), *The Great Chain of Being* and *Italian Phenomenology.* 1981 ISBN 90-277-1071-6
12. Tymieniecka, A-T. (ed.), *The Philosophical Reflection of Man in Literature.* Selected Papers from Several Conferences held by the International Society for Phenomenology and Literature in Cambridge, Massachusetts. Includes the essay by A-T. Tymieniecka, *Poetica Nova.* 1982 ISBN 90-277-1312-X
13. Kaelin, E. F., *The Unhappy Consciousness.* The Poetic Plight of Samuel Beckett. An Inquiry at the Intersection of Phenomenology and literature. 1981 ISBN 90-277-1313-8
14. Tymieniecka, A-T. (ed.), *The Phenomenology of Man and of the Human Condition.* Individualisation of Nature and the Human Being. (Part I:) Plotting the Territory for Interdisciplinary Communication. 1983 *Part II* see below under Volume 21. ISBN 90-277-1447-9

Analecta Husserliana

15. Tymieniecka, A-T. and Calvin O. Schrag (eds.), *Foundations of Morality, Human Rights, and the Human Sciences.* Phenomenology in a Foundational Dialogue with Human Sciences. 1983 ISBN 90-277-1453-3
16. Tymieniecka, A-T. (ed.), *Soul and Body in Husserlian Phenomenology.* Man and Nature. 1983 ISBN 90-277-1518-1
17. Tymieniecka, A-T. (ed.), *Phenomenology of Life in a Dialogue Between Chinese and Occidental Philosophy.* 1984 ISBN 90-277-1620-X
18. Tymieniecka, A-T. (ed.), *The Existential Coordinates of the Human Condition: Poetic – Epic – Tragic.* The Literary Genre. 1984 ISBN 90-277-1702-8
19. Tymieniecka, A-T. (ed.), *Poetics of the Elements in the Human Condition.* (Part 1:) The Sea. From Elemental Stirrings to Symbolic Inspiration, Language, and Life-Significance in Literary Interpretation and Theory. 1985
For Part 2 and 3 *see below* under Volumes 23 and 28. ISBN 90-277-1906-3
20. Tymieniecka, A-T. (ed.), *The Moral Sense in the Communal Significance of Life.* Investigations in Phenomenological Praxeology: Psychiatric Therapeutics, Medical Ethics and Social Praxis within the Life- and Communal World. 1986 ISBN 90-277-2085-1
21. Tymieniecka, A-T. (ed.), *The Phenomenology of Man and of the Human Condition.* Part II: The Meeting Point Between Occidental and Oriental Philosophies. 1986 ISBN 90-277-2185-8
22. Tymieniecka, A-T. (ed.), *Morality within the Life- and Social World.* Interdisciplinary Phenomenology of the Authentic Life in the "Moral Sense'. 1987
Sequel to Volumes 15 and 20. ISBN 90-277-2411-3
23. Tymieniecka, A-T. (ed.), *Poetics of the Elements in the Human Condition.* Part 2: The Airy Elements in Poetic Imagination. Breath, Breeze, Wind, Tempest, Thunder, Snow, Flame, Fire, Volcano . . . 1988 ISBN 90-277-2569-1
24. Tymieniecka, A-T., *Logos and Life.* Book I: Creative Experience and the Critique of Reason. 1988 ISBN Hb 90-277-2539-X; Pb 90-277-2540-3
25. Tymieniecka, A-T., *Logos and Life.* Book II: The Three Movements of the Soul. 1988 ISBN Hb 90-277-2556-X; Pb 90-277-2557-8
26. Kaelin, E. F. and Calvin O. Schrag (eds.), *American Phenomenology.* Origins and Developments. 1989 ISBN 90-277-2690-6
27. Tymieniecka, A-T. (ed.), *Man within his Life-World.* Contributions to Phenomenology by Scholars from East-Central Europe. 1989 ISBN 90-277-2767-8
28. Tymieniecka, A-T. (ed.), *The Elemental Passions of the Soul.* Poetics of the Elements in the Human Condition, Part 3. 1990 ISBN 0-7923-0180-3
29. Tymieniecka, A-T. (ed.), *Man's Self-Interpretation-in-Existence.* Phenomenology and Philosophy of Life. – Introducing the Spanish Perspective. 1990 ISBN 0-7923-0324-5
30. Rudnick, H. H. (ed.), *Ingardeniana II.* New Studies in the Philosophy of Roman Ingarden. With a New International Ingarden Bibliography. 1990 ISBN 0-7923-0627-9

Analecta Husserliana

31. Tymieniecka, A-T. (ed.), *The Moral Sense and Its Foundational Significance: Self, Person, Historicity, Community.* Phenomenological Praxeology and Psychiatry. 1990 ISBN 0-7923-0678-3
32. Kronegger, M. (ed.), *Phenomenology and Aesthetics.* Approaches to Comparative Literature and Other Arts. Homages to A-T. Tymieniecka. 1991 ISBN 0-7923-0738-0
33. Tymieniecka, A-T. (ed.), *Ingardeniana III.* Roman Ingarden's Aesthetics in a New Key and the Independent Approaches of Others: The Performing Arts, the Fine Arts, and Literature. 1991
Sequel to Volumes 4 and 30 ISBN 0-7923-1014-4
34. Tymieniecka, A-T. (ed.), *The Turning Points of the New Phenomenological Era.* Husserl Research – Drawing upon the Full Extent of His Development. 1991 ISBN 0-7923-1134-5
35. Tymieniecka, A-T. (ed.), *Husserlian Phenomenology in a New Key.* Intersubjectivity, Ethos, the Societal Sphere, Human Encounter, Pathos. 1991 ISBN 0-7923-1146-9
36. Tymieniecka, A-T. (ed.), *Husserl's Legacy in Phenomenological Philosophies.* New Approaches to Reason, Language, Hermeneutics, the Human Condition. 1991 ISBN 0-7923-1178-7
37. Tymieniecka, A-T. (ed.), *New Queries in Aesthetics and Metaphysics.* Time, Historicity, Art, Culture, Metaphysics, the Transnatural. 1991 ISBN 0-7923-1195-7
38. Tymieniecka, A-T. (ed.), *The Elemental Dialectic of Light and Darkness.* The Passions of the Soul in the Onto-Poiesis of Life. 1992 ISBN 0-7923-1601-0
39. Tymieniecka, A-T. (ed.), *Reason, Life, Culture, Part I.* Phenomenology in the Baltics. 1993 ISBN 0-7923-1902-8
40. Tymieniecka, A-T. (ed.), *Manifestations of Reason: Life, Historicity, Culture.* Reason, Life, Culture, Part II. Phenomenology in the Adriatic Countries. 1993 ISBN 0-7923-2215-0
41. Tymieniecka, A-T. (ed.), *Allegory Revisited.* Ideals of Mankind. 1994 ISBN 0-7923-2312-2
42. Kronegger, M. and Tymieniecka, A-T. (eds.), *Allegory Old and New.* In Literature, the Fine Arts, Music and Theatre, and Its Continuity in Culture. 1994 ISBN 0-7923-2348-3
43. Tymieniecka, A-T. (ed.): *From the Sacred to the Divine.* A New Phenomenological Approach. 1994 ISBN 0-7923-2690-3
44. Tymieniecka, A-T. (ed.): *The Elemental Passion for Place in the Ontopoiesis of Life.* Passions of the Soul in the *Imaginatio Creatrix.* 1995 ISBN 0-7923-2749-7
45. Zhai, Z.: *The Radical Choice and Moral Theory.* Through Communicative Argumentation to Phenomenological Subjectivity. 1994 ISBN 0-7923-2891-4
46. Tymieniecka, A-T. (ed.): *The Logic of the Living Present.* Experience, Ordering, Onto-Poiesis of Culture. 1995 ISBN 0-7923-2930-9

Analecta Husserliana

47. Tymieniecka, A-T. (ed.): *Heaven, Earth, and In-Between in the Harmony of Life.* Phenomenology in the Continuing Oriental/Occidental Dialogue. 1995 ISBN 0-7923-3373-X
48. Tymieniecka, A-T. (ed.): *Life. In the Glory of its Radiating Manifestations.* 25th Anniversary Publication. Book I. 1996 ISBN 0-7923-3825-1
49. Kronegger, M. and Tymieniecka, A-T. (eds.): *Life. The Human Quest for an Ideal.* 25th Anniversary Publication. Book II. 1996 ISBN 0-7923-3826-X
50. Tymieniecka, A-T. (ed.): *Life. Phenomenology of Life as the Starting Point of Philosophy.* 25th Anniversary Publication. Book III. 1997 ISBN 0-7923-4126-0
51. Tymieniecka, A-T. (ed.): *Passion for Place. Part II.* Between the Vital Spacing and the Creative Horizons of Fulfilment. 1997 ISBN 0-7923-4146-5
52. Tymieniecka, A-T. (ed.): *Phenomenology of Life and the Human Creative Condition.* Laying Down the Cornerstones of the Field. Book I. 1997 ISBN 0-7923-4445-6
53. Tymieniecka, A-T. (ed.): *The Reincarnating Mind, or the Ontopoietic Outburst in Creative Virtualities.* Harmonisations and Attunement in Cognition, the Fine Arts, Literature. Phenomenology of Life and the Human Creative Condition. Book II. 1997 ISBN 0-7923-4461-8
54. Tymieniecka, A-T. (ed.): *Ontopoietic Expansion in Human Self-Interpretation-in-Existence.* The I and the Other in their Creative Spacing of the Societal Circuits of Life. Phenomenology of Life and the Creative Condition. Book III. 1997 ISBN 0-7923-4462-6
55. Tymieniecka, A-T. (ed.): *Creative Virtualities in Human Self-Interpretation-in-Culture.* Phenomenology of Life and the Human Creative Condition. Book IV. 1997 ISBN 0-7923-4545-2
56. Tymieniecka, A-T. (ed.): *Enjoyment.* From Laughter to Delight in Philosophy, Literature, the Fine Arts and Aesthetics. 1998 ISBN 0-7923-4677-7
57. Kronegger M. and Tymieniecka, A-T. (eds.): *Life. Differentiation and Harmony... Vegetal, Animal, Human.* 1998 ISBN 0-7923-4887-7
58. Tymieniecka, A-T. and Matsuba, S. (eds.): *Immersing in the Concrete.* Maurice Merleau-Ponty in the Japanese Perspective. 1998 ISBN 0-7923-5093-6
59. Tymieniecka, A-T. (ed.): *Life - Scientific Philosophy/Phenomenology of Life and the Sciences of Life.* Ontopoiesis of Life and the Human Creative Condition. 1998 ISBN 0-7923-5141-X
60. Tymieniecka, A-T. (eds.): *Life - The Outburst of Life in the Human Sphere.* Scientific Philosophy / Phenomenology of Life and the Sciences of Life. Book II. 1998 ISBN 0-7923-5142-8
61. Tymieniecka, A-T. (ed.): *The Aesthetic Discourse of the Arts.* Breaking the Barriers. 2000 ISBN 0-7923-6006-0
62. Tymieniecka, A-T. (ed.): *Creative Mimesis of Emotion.* From Sorrow to Elation; Elegiac Virtuosity in Literature. 2000 ISBN 0-7923-6007-9

Analecta Husserliana

63. Kronegger, M. (ed).: *The Orchestration of The Arts – A Creative Symbiosis of Existential Powers*. The Vibrating Interplay of Sound, Color, Image, Gesture, Movement, Rhythm, Fragrance, Word, Touch. 2000 ISBN 0-7923-6008-7
64. Tymieniecka, A-T. and Z. Zalewski (eds.): *Life - The Human Being Between Life and Death*. A Dialogue Between Medicine and Philosophy, Recurrent Issues and New Approaches. 2000 ISBN 0-7923-5962-3

Kluwer Academic Publishers – Dordrecht / Boston / London

GPSR Compliance
The European Union's (EU) General Product Safety Regulation (GPSR) is a set of rules that requires consumer products to be safe and our obligations to ensure this.

If you have any concerns about our products, you can contact us on

ProductSafety@springernature.com

In case Publisher is established outside the EU, the EU authorized representative is:

Springer Nature Customer Service Center GmbH
Europaplatz 3
69115 Heidelberg, Germany

www.ingramcontent.com/pod-product-compliance
Ingram Content Group UK Ltd.
Pitfield, Milton Keynes, MK11 3LW, UK
UKHW020135160926
475607UK00003B/50

* 9 7 8 0 7 9 2 3 5 9 6 2 3 *